Conservation of Historic Buildings and their Contents

Addressing the Conflicts

Edited by David Watt and Belinda Colston

Routledge
Taylor & Francis Group

LONDON AND NEW YORK

First published 2003 by Donhead Publishing Ltd

Published 2015 by Routledge
2 Park Square, Milton Park, Abingdon, Oxon OX14 4RN
711 Third Avenue, New York, NY 10017, USA

Routledge is an imprint of the Taylor & Francis Group, an informa business

Copyright © Taylor & Francis 2003
In association with De Montfort University

ISBN 13: 978-1-873394-63-2 (pbk)

Contents

Winter Smoking Room, Cardiff Castle (see page 81) (Cardiff County Council).

Preface

The eight papers and discussion sessions of a conference entitled 'Where Conservation Meets Conservation': The Interface between Historic Buildings and their Contents, held at De Montfort University on 9 September 2002, form the basis of these published proceedings. The papers, in themselves, provide an important commentary on the development and practice of conservation, both of historic buildings and their contents. Taken together, the value that comes from the shared knowledge and experience of the authors is immeasurable.

Where Does Conservation Meet Conservation?

DAVID WATT AND BELINDA COLSTON

The title of the conference held at De Montfort University in Leicester on 9 September 2002 was '*Where Conservation Meets Conservation*': *The Interface between Historic Buildings and their Contents*. Although altered for the purpose of these proceedings, the original title reflects the dichotomy that is too often present in planning and implementing works within the historic cultural environment.

De Montfort University has offered post-graduate training in architectural conservation, first as a Diploma and latterly at Master's level, since 1974 and has seen at first hand the changing needs and expectations of its students. The programme running today is different in many ways from its predecessor and now offers a greater breadth, and challenge, in coming to terms with subjects such as material science and landscape history. Hopefully, those graduating from the university will have a more rounded appreciation of architectural conservation and have the ability to enter into informed dialogue with other members of an interdisciplinary project team.

More recently, De Montfort University has offered a MSc programme in conservation science, which now runs in a distance-learning format. This route allows students from different countries to obtain a theoretical grounding in the various aspects of this growing subject, backed up with practical sessions at an annual summer school. As with the architectural conservation programme, emphasis is placed on the interdisciplinary nature of the subject and students are encouraged to see their work in the context of the wider historic cultural environment.

The study of conservation also offers students a deeper intellectual challenge that can be pursued at MPhil or PhD level. Current research students are engaged in projects ranging from a study of traditional stone slate roofing techniques and the effects of hydrocarbon pollution on the underground parts of ancient stone monuments to the investigation of the binding media used in post-Byzantine icons and the current health risks associated with chemical residues from past conservation treatments of historic textile and herbarium collections.

Figure 1 *Delegates from the Archaeological Survey of India during a practical lime day with Sir Bernard Feilden at Stiffkey Old Hall, Norfolk, in May 1996. (David Watt)*

There have also been opportunities to bring together the various strands of conservation education through international training and consultancy, such as the 'Structural and Material Conservation' programme provided for selected delegates from the Archaeological Survey of India during the 1990s (Figure 1).

Aside from academic activities, De Montfort University also established and continues an active involvement with the *Journal of Architectural Conservation*. Published by Donhead Publishing since 1995, this international publication is concerned with the conservation of historic buildings, monuments, places, and landscapes; many of its papers deal with the interrelated theme of the building and interior finishes.

The enduring commitment of De Montfort University to teaching and learning in conservation places it in a strong position to promote a broader exchange of ideas and values across the key conservation disciplines. This is the basis on which the idea of the 'Where Conservation Meets Conservation' conference, and indeed the title of the event, was founded.

It was not so many years ago that a colloquium entitled *Conservation Research: Needs and Provision* was held at De Montfort University with the aim of providing:

... a reflective pause in which representatives of the various heritage professions, who either need or provide conservation research, could look back over the period since the last stock-taking and consider an agenda for the future'.[1]

Much has changed during the intervening six years, both in academic institutions and professional practice, and there seems even less time to pause and reflect on what is going on around us. We have, with 'Where Conservation Meets Conservation', looked backward to easier and, some would say, happier times and forward to the challenges that await us. The story is told through the papers in this publication, for which the editors would like to thank the authors, sponsors, and all concerned in managing what was a successful and enjoyable day.

The papers

The stage was set with a fascinating and thought-provoking paper by Dr Nicholas Stanley-Price entitled Movable:Immovable – A Historic Distinction and its Consequences. In his paper, the Director-General of ICCROM considers the different fields of specialization that have grown up and the need for greater integration in conservation policies and practice.

Following naturally on from this opening paper comes Beyond the Divide – Experiences from Scottish Conservation by Carol Brown of Historic Scotland. In this, the author covers the birth of the distinctive integrated approach to Scottish conservation stemming from the 1960s, through the formation of Historic Scotland's Technical Conservation Research and Education Division and the Scottish Conservation Bureau, and the current role of the Scottish Conservation Forum in Training and Education as a point of contact and focus for finding common ground.

Moving on to issues of sustainability and the paper entitled 'Places' and 'Stuff': Is it Only the Language of Conservation that is Changing?, May Cassar considers the wider context within which conservation of the material heritage is practised and argues a case for conservation being the management of a non-renewable resource. A key point of reference coming from this paper is the commonality that can be used to demonstrate the relevance of the historic environment to all members of society and the priorities that should be set to ensure its place in our future.

How to balance the needs of sensitive decorative layers with those of a congregation can be a frustrating business. In his paper, Working Buildings: The Effect of Building Use on the Conservation of Wall Paintings and Polychrome Surfaces, Tobit Curteis considers the differing requirements of the building and its users, and draws attention to the role of the conservator in advising on the ways in which conditions can be achieved that are acceptable to the people using the building, as well as being suitable for the conservation of the building fabric and the objects displayed within it.

The Fascination of White Gloves by Michael Morrison, partner with Purcell Miller Tritton Architects, provided an entertaining, yet important, practitioner's view of working with historic buildings. It has not been possible for Michael to provide the text from his presentation, although, as editors, who could want a more compelling explanation?

> When last in contact with him [Michael] ... he doubted he would have time to look at his notes ... as power and conditions were difficult. Since then he has encountered the worst blizzard in the Antarctic for some considerable years which ripped their tent apart and left them stranded for four days awaiting a helicopter to rescue them. This didn't materialize so they had to walk across the ice to catch the ice-breaker en route to Cape Adare! All jolly exciting stuff in retrospect I'm sure.[2]

The abstract submitted by Michael Morrison, together with his input into the discussion sessions, will hopefully provide a record of his views on 'where conservation meets conservation':

> Architects carrying out conservation work have a tendency to place undue weight on specific bits of scientific data. It is much easier to define the needs of a single object than of a complex multi-functional system. In the (necessary) move over the last 20 years to make conservation techniques more soundly based in science there has been a tendency to lose sight of the bigger picture – the 'Spirit of the Place' in National Trust terms. In the extreme, the measures to preserve the contents of a building may be destroying the fabric of the building itself. The impact of environmental control systems on the fabric of any building needs to be evaluated in the context of the life expectancy of the system and the implications of renewal. Why is a change to part of the building fabric more acceptable than a change to the objects in the building?[3]

Staying with The National Trust, and as a counterbalance to the views of Michael Morrison as project architect, Sarah Staniforth and Katy Lithgow responded with their joint paper, *When Conservator Meets Architect and Engineer*. In this, changes to external climate and internal environmental conditions are considered in relation to both the building and associated collections. The use of conservation heating systems to create constant humidity is one answer to the problems faced in unheated properties. The introduction of such systems and other preventive conservation measures can, in themselves, pose significant risks to collections and interiors through damage, dust, and theft. The manner in which these risks are controlled, through forward planning and good project management, can make the difference between success and failure. This paper therefore provides much useful guidance for all members of a project team.

The challenges of conserving, whilst at the same time using, a historic building are legend. Being faced with an irate client who cannot understand why his bed is surrounded by Acrow® props during works on the floor above or who wishes to host impromptu dinner parties in the middle of contract works can be trying at the best of times. There are lessons to be learnt from each and every situation. The work currently being undertaken at Cardiff Castle and presented by John Edwards in *Conserving Cardiff Castle – Planning for Success* provides a model for how complex, interdisciplinary projects can and should be managed. In this, the importance of allowing for (and adequately resourcing) a programme of research and analysis is made clear. Drawing together and resolving technical and philosophical issues in the context of Cardiff Castle offers a challenge, but with it hoped-for answers to mutual problems.

Finally, and as a fitting close to the conference and these proceedings, Donald Hankey provides a powerful argument for better management of conservation, whether in relation to buildings or their contents. Based on extensive personal involvement in international conservation projects, Donald Hankey in his paper *Management of the Historic Environment – The Broad Nature of the Process* makes clear the need for consensus between all stakeholders. This, in his words, can only be achieved 'by promoting the best scientific, social, and cultural understanding'. This must surely be the way forward.

Conclusion

And so, finally, where does conservation meet conservation? The purpose of the conference, as expressed in the sub-title, was to consider the interface between historic buildings and their contents. This has been done and along the way we have considered the various layers of involvement and interaction that go to make what is an increasingly complex, interdisciplinary activity (Figure 2).

From the intricacies of micro-conservation as applied to medieval wall paintings to the macro-conservation level of addressing the problems of whole cities, conservation meets conservation at many levels.

Figure 2 *A collection of gas meters and other related appliances at the Fakenham Museum of Gas and Local History in Norfolk (courtesy of the Fakenham Gasworks Museum Trust). The gasworks are scheduled as an ancient monument, based around a number of nineteenth-century buildings housing the machinery used in the former production of town gas. In conservation terms, many of the relationships between the site, buildings, machinery, and collections are complex and a clear example, in an industrial context, of where conservation meets conservation.*[4] *(David Watt)*

What has come from the conference is a challenge to all that work, on the one hand, with historic buildings and, on the other, with collections and contents. Bridges must be built between the various disciplines and a closer dialogue entered into at each and every opportunity.

As our students are often told, 'conservation is easy – you just say 'no'. What is hard is to say 'yes' and make a difference.'

Biography

David Watt BSc (Hons), Dip Arch Cons (Leic), PhD, MSc, FRICS, IHBC
Dr David Watt is a Chartered Building Surveyor and Senior Research Fellow at De Montfort University, Leicester. He promotes, conducts, and publishes research on various aspects of architectural conservation and building pathology, and is particularly interested in the use of traditional building materials and the influence of people and environmental conditions on buildings and monuments. Dr David Watt is editor of the *Journal of Architectural Conservation*.

Belinda Colston BSc (Hons), PhD, CChem, MRSC
Dr Belinda Colston is a Senior Lecturer in the School of Molecular Sciences at De Montfort University, Leicester. She is engaged in a range of research and consultancy projects concerning aspects of stone decay and environmental impact on sensitive collections. Dr Belinda Colston is Programme Leader for the MSc Conservation Science course at De Montfort University and adviser to the ICCROM 'CURRIC' programme.

Notes

1 Foley, K., and Shacklock, V., *Conservation Research: Needs and Provision*, proceedings of a colloquium held on 1 July 1997 at De Montfort University, De Montfort University, Leicester (1997).
2 Cooper, L., personal communication with David Watt, 28 January 2003.
3 Morrison, M., abstract submitted for paper entitled 'The Fascination of White Gloves', 3 March 2002.
4 For further information relating to Fakenham Gasworks, see: Watt, D., 'Fakenham Gasworks: Over 150 Years of History', *Transactions of Ancient Monument Society*, Vol 43, 1999, pp. 25–44.

Movable:Immovable – A Historic Distinction and its Consequences

NICHOLAS STANLEY-PRICE

Abstract

The challenge to preserve both a historic building and its contents is but one example of the need to overcome traditional distinctions between immovable and movable property conservation. Nevertheless, historically they have been considered different fields of specialization, each with its own educational programmes, practitioners, and professional organizations.

The paper reviews the movable:immovable distinction in the light of the origins of the international organizations ICCROM (International Centre for the Study of the Preservation and Restoration of Cultural Property), ICOM (International Council of Museums), ICOMOS (International Council on Monuments and Sites), and IIC (International Institute for the Conservation of Historic and Artistic Works). It advocates greater integration of the two fields in conservation policies.

Introduction

'Where conservation meets conservation'. To what extent is it true that conservation of historic buildings is a distinct field of practice from conservation of objects found inside those buildings? Is there sometimes a grey area between the two?

If there is, it is due to a number of reasons. These would include the differences in the professional education of architects and object conservators – differences that concern the qualifications needed to practice, the scale of work that each type of practitioner prefers, and the relative scope for creative re-use of heritage (usually the greater in the case of buildings).

All of these differences are very real. But the existence of this distinction (which is a real one) owes much to the way that conservation has developed historically. That the phrase 'where conservation meets conservation' is not meaningless can be attributed, in part, to the history of the conservation field. Much national practice has been influenced, for better or worse, by international example.

This paper considers some aspects of the development of international conservation organizations. This will be by reference to a distinction that is often made in conservation – the distinction between movable and immovable. For example, historic buildings are usually viewed as immovable, and their contents as movable (with the exception of those items defined in English law as 'fixtures'). What are the origins of this distinction within conservation?

I will argue that it is, to some extent, responsible for barriers between different areas of professional practice today. If we wish to discourage this distinction and integrate the best of both areas of practice, it is important to understand how it originated.

The movable:immovable distinction

The movable:immovable terminology is widespread in international conservation texts. What is its origin? I assume it is derived from the past use of the term 'cultural property' to describe in general terms what we aim to conserve. The term 'cultural property' was first used in English in a legal context in the Hague Convention of 1954, and was used subsequently for the Convention on the Means of Prohibiting and Preventing the Illicit Import, Export and Transfer of Ownership of Cultural Property of 1970.[1] It also features in the title of the specialized conservation centre (the 'Rome Centre') that was founded in Rome in the late 1950s and that is now known for short as ICCROM (International Centre for the Study of the Preservation and Restoration of Cultural Property).

In much legislation deriving from the Civil Law, a distinction is made between movable and immovable property. This is not the case under the Common Law that is applied in the United Kingdom, though the notion of property itself is of fundamental importance.[2] The early conventions and recommendations of UNESCO (United Nations Educational, Scientific and Cultural Organization) refer to the protection of cultural property. All of the texts up to and including the 1970 Convention refer to movable and immovable property.

Since then, the notion of property has come to be seen as less appropriate than that of 'heritage'.[3] Whereas the term 'property' conveys ideas of ownership rights and of commercial value, 'heritage' implies rather a legacy to inherit and to pass on as a social value to future generations. The term 'heritage' grew increasingly popular in the 1960s. It first came to be used in English for an international agreement in the European

Convention on the Protection of the Archaeological Heritage of 1969 (the London Convention); and then, significantly, in UNESCO's 1972 Convention concerning the Protection of the World Cultural and Natural Heritage.

With the shift from the term 'cultural property' to 'cultural heritage', the tendency to distinguish formally between movable and immovable property was reduced. But the terminology is still widely used, especially in the Romance languages. It is found in legislation, titles, and much written commentary.

National legislation often refers to cultural material as, for example, monuments, antiquities, and works of art. None of these terms denote exclusively either movable or immovable items. Even monuments can fall into either category. In England, a monument is viewed unmistakably as an immovable. But in many French-, German- and Spanish-language legislations, the term 'monument' can refer to both immovables and movables. The same is true even in English-inspired legislation in several former British colonies.[4]

Apart from these terminological problems, the difficulties with the movable:immovable distinction are well-known. They concern heritage items that seem not to fit the distinction at all, those that could fit into either category, or, worse, those that change from one category to the other.

Items normally considered to be immovable can sometimes be moved, whether legally for their protection and conservation, or illegally for sale. Buildings, machinery, and industrial installations are moved to open-air museums in order to save them; or, what would be considered fixtures in a building are detached and removed for their better conservation. Examples would include mosaics, wall paintings, sculptured reliefs and stelae found on archaeological sites. When detachment was considered the only way of conserving such fixtures, they were treated as movables to be conserved and displayed in museums. As conservation techniques improved, and policies emphasized conservation *in situ*, such fixtures became regarded as immovables. A shipwreck provides the converse case, in which what was originally a movable object (a ship) becomes considered a site with remains usually to be preserved *in situ*.

Historic houses and their contents epitomize the hazards of the distinction. Successful conservation requires a fusion between two different areas of expertise: namely, the conservator of movables (often referred to as the objects conservator), and the conservator of the immovable (often referred to as the architectural conservator or the conservation architect). Much of the formal education and many of the professional associations to which the specialists might belong reflect the movable:immovable distinction. How has this come about?

To a certain extent the distinction is reflected in the origins of some of the international organizations. The following account represents a preliminary series of observations, as much of the relevant information is historical and hard to locate. It also concentrates on the built heritage and museums, with only scant reference to libraries and archives.

Development of the international organizations

The greater part of international conservation activity takes place under the auspices of a number of organizations that were founded 40–50 years ago. The nature of the work that they facilitate through professional exchanges owes much to their founding goals. Two of those organizations are international NGOs (non-governmental organizations), namely ICOM (International Council of Museums) and ICOMOS (International Council on Monuments and Sites). They seem at first sight to reflect the movable:immovable distinction – ICOM is concerned with museums and ICOMOS with monuments and sites. But it has not always been as simple as that, as the following account attempts to demonstrate.

From the International Museums Office to ICOM

The first steps towards international co-operation in the cultural conservation field are to be found in the work of the International Museums Office (IMO). This was established in 1927 by the Institute of Intellectual Cooperation, in turn created by the League of Nations.[5] During the 1930s, the IMO helped organize a series of international conferences whose results laid the foundations for future developments. Not only did they help create international networks of specialists, but their conclusions and recommendations formed the basis of future charters enunciating conservation principles.

The first such conference was the one held in Rome in 1930, on the study of scientific methods for the examination and preservation of works of art.[6] Concern expressed at the meeting over the restoration of paintings and the use of varnish led to a working group and eventually a *Manual on the Conservation of Paintings*.[7]

At the Rome conference, the IMO announced that another conference would be held the following year in Athens, on the conservation of archaeological monuments.[8] This meeting gave rise to the *Charter of Athens* (1931), the first such document proposing an international consensus on the principles of monument restoration.

Within the archaeological field, the problem of the regulation of excavations led to the Cairo Conference of 1937, organized by the IMO with the Egyptian Government. The International Conference on Excavations had two important outcomes: the publication of a *Manual on the Technique of Archaeological Excavations* and a series of recommendations for the conduct of excavations and international collaboration, which were duly adopted by the League of Nations.[9]

Many of the initiatives of the IMO were taken up after the Second World War by ICOM, founded as an international NGO in 1946. It quickly established co-operation with the new intergovernmental organization UNESCO, which came into being the same year.

The definition of a museum used in the Founding Resolutions of ICOM places the emphasis on collections:

> The word 'Museums' includes all collections, open to the public, of artistic, technical, scientific, historical or archaeological material, including zoos and botanical gardens, but excluding libraries, except in so far as they maintain permanent exhibition rooms.[10]

At ICOM's first general conference in 1948, twelve permanent international committees were set up.[11] Although the primary object of interest for ICOM was collections, several of the international committees were concerned with museum-type functions performed at non-museum institutions. These included zoological gardens, aquaria and botanical gardens, national parks and nature reserves, and museums of archaeology and history, and historical sites. The broad range of institutions recognized to carry out museum-type activities was formally acknowledged in a revised definition of 'Museums' that was adopted for the ICOM Statutes in 1961:

> Article 4. Within this definition fall:
> a. exhibition galleries permanently maintained by public libraries and collections of archives,
> b. historical monuments and parts of historical monuments or their dependencies, such as cathedral treasuries, historical, archaeological and natural sites, which are officially open to the public,
> c. botanical and zoological gardens, aquaria, vivaria, and other institutions which display living specimens,
> d. natural reserves.[12]

At the first ICOM general conference, conservation also received attention. Concern over the cleaning of paintings led to what was to become the ICOM Commission on the Care of Paintings.[13] This move coincided in time with the plans, already far advanced, to establish an independent institute devoted to conservation and restoration. This was the institute now known as the International Institute for the Conservation of Historic and Artistic Works (IIC).

It is interesting to note the evolution of suggestions for the name of this Institute. Perhaps because its main proponents resided in countries following Common Law systems, its title made no reference to cultural property, nor to movable and immovable distinctions. The original proposal in 1947 was for an 'Association for the Conservation of Cultural Holdings'. This encountered objections to both the 'holdings' (as being untranslatable) and to the 'cultural' (because of the adverse connotations of 'culture' at the time). Following an alternative proposal for 'The International Institute for the Conservation of Objects of Art and Archaeology', it was eventually incorporated under English law in London in 1950 as the International Institute for the Conservation of Museum Objects.[14]

The area of interest implied by the title of the new Institute was broad compared with the contemporary ICOM Commission for Care of Paintings (founded in 1948). But the new Institute had an uneasy relationship with the Commission.[15] In fact, as time passed, the Institute saw its work as being even broader than museum objects alone. Eventually, in 1959, it changed its name to the current title, the International Institute for the Conservation of Historic and Artistic Works. The aim was to demonstrate that its interests included buildings and were not limited to museum objects.[16]

Sites and monuments: a parallel development

By 1950, therefore, there existed two international organizations that brought together specialists in conservation. The IIC was established as an association of professional conservators, and invited individuals and institutes to join. ICOM, in turn, consisted of a federation of national committees of individuals devoted to museum work, of which conservation represented, of course, only one function. Within this function, the care of paintings was the subject of the first specialist commission created in 1948, followed in 1951 by one devoted to museum laboratories.[17]

ICOM's definition of museums, as confirmed in the revised statutes of 1961, included archaeological and historic sites and monuments. Indeed, ICOM was often consulted by UNESCO on questions concerning sites. For any location that had collections, whatever the nature of the repository (archives, sites, monuments, parks, etc), ICOM was viewed as the appropriate specialized body. But the ethical and technical problems of the restoration of monuments were recognized to be a different field of expertise from that of ICOM, and one that required its own specialist network. The extensive damage caused to monuments by the Second World War had made this an urgent need.

In 1949, UNESCO formed an expert working group on the protection of monuments, which in turn recommended the creation of a permanent international committee. This took shape in 1951 in the International Committee on Monuments, Artistic and Historical Sites, and Archaeological Excavations.[18] The Committee and ICOM each had its own domain, but close liaison was essential as there were subjects of common interest. Its Director, G.H. Rivière, declared that 'ICOM had from the outset been in favour of the creation of a separate body to consider questions in the field of monuments and archaeological excavations'.[19] Such a body had now been created.

In addition to the needs of war-damaged monuments, the Committee concerned itself with another problem inherited from before the War, namely the conduct of archaeological excavations and the finds derived from them. ICOM had been asked by UNESCO to follow up on implementation of the recommendations of the 1937 Cairo Conference on Archaeological Excavations, which had become largely ignored following the outbreak of war. ICOM's newly-established International Subject Committee on Museums of Archaeology and History met in Naples and Pompeii in 1953. Its own conclusions were taken up by the UNESCO International Committee on Monuments, which in turn produced a text that was subsequently approved as a UNESCO Recommendation (the Recommendation on International Principles applicable to Archaeological Excavations, adopted at the General Conference of UNESCO in 1956).[20]

In the same year of 1953, a sub-committee of the International Committee on Monuments, chaired by the Director of ICOM, recommended the establishment of a specialized centre for conservation. This was formally approved in 1956 as the International Centre for the Study of the Preservation and Restoration of Cultural Property.[21] Significantly, from its foundation, its mandate covered all forms of cultural property, movable and immovable.

The new Centre (now known as ICCROM) was founded as an intergovernmental organization. It worked closely with the non-governmental organization, ICOM, especially on the conservation of museum collections. Their collaboration resulted, for example, in a series of publications under a Rome Centre–ICOM imprint.

ICOM continued during the 1950s to play an important advisory role to UNESCO on archaeological matters. For instance, it worked with UNESCO to launch, in 1960, the archaeological salvage campaigns in Nubia. But, by then, moves were afoot to create a specialized NGO for monuments and sites.

In 1957, the French association of restoration architects organized, with the help of UNESCO, an International Congress of Architects and Technicians of Historic Monuments. The conclusions of the Athens conference of 1931 were taken up for further discussion. A resolution was addressed to UNESCO to create an international association that would provide a link between restoration architects and technicians.[22] The second congress, held under UNESCO's auspices in Venice in 1964, took the decision to do so and ICOMOS was founded as an international NGO in the following year (1965). The founders of ICOMOS adopted the model and structure of ICOM, a kind of federation of national committees.[23] It also adopted as its fundamental doctrinal document the *Venice Charter*, which had been approved at the congress of the previous year.

That ICOMOS was concerned with the conservation and restoration of monuments and sites and that collections, even when found in 'monuments', were considered the province of ICOM is reinforced in the wording of its statutes adopted in 1978. These specifically excluded as part of ICOMOS' area of interest:

> Museum collections housed in monuments; archaeological collections preserved in museums or housed in museums or exhibited at archaeological or historic site museums; or open-air museums.[24]

These were evidently recognized as falling into ICOM's sphere of responsibility.

Specialized committees of ICOM and ICOMOS

Right from their foundations, ICOM and ICOMOS created sub-groups of members having common interests. Variously called International Committees or Commissions, they brought together specialists who organized their own meetings. In 1967, the two ICOM Committees of Scientific Museum Laboratories and Care of Paintings (as it had been re-named in 1949) merged to form the International Committee for Conservation. Since then, over 20 Working Groups of the Committee have been founded, bringing together conservators with similar interests.

Among the many Committees and Working Groups of ICOM and ICOMOS that exist today – some more active than others – there exists a certain overlap of interests. Table 1 shows a selection of them, with their foundation dates where known (this information is unevenly published).

As Table 1 shows, all three organizations have specialized groups concerned with professional training, and with documentation. Within documentation, the Architectural Photogrammetry Committee of ICOMOS itself has expert groups on, for example, Archaeological Conservation and Museum Objects, which in turn include working groups on such topics as archaeological management, rock art, and objects. Particular materials, such as wood, stone, and glass, are of interest to all three organizations. Legal issues have received more attention in recent years in this field as in many others. The ICOM-CC (International Council of Museums – Conservation Committee) and ICOMOS have recently founded specialized groups to consider legal issues, which also form part of the area of interest of ICOM's International Committee on Management.

Disaster preparedness is a theme common to all three, being a focus of interest within the ICOM Committee on Museum Security and in the ICOM-CC working group on preventive conservation. Finally, with regard to the theme of this publication, historic houses are precisely the focus of the International Committee of ICOM for Historic House Museums. Its interests include such topics as management and security of historic houses that are now museums.

Overlap between organizations

As early as 1963 there was concern expressed at the overlapping between the different organizations and moves were made to develop closer links between the IIC, ICOM, and the Rome Centre (as ICCROM was then known).[25] Subsequent years saw some attempts to do so: for example, in the 1960s, there was an International Committee for the Coordination of Publications on Conservation, and a Joint Stone Committee.[26] On both of these was represented ICOMOS following its founding in 1965. In 1970, the IIC organized a meeting with ICOM and the Rome Centre

Table 1 (right) Selected International Committees and Working Groups of ICOM and ICOMOS. Foundation dates are given where known. Sources:
http://icom.museum/internationals.html
http://www.icom-cc.org/index/organiz/hpagewg.htm (no longer active)
http://www.international.icomos.org/address.htm
(August 2002), published references (see text), and personal communications.

ICOM (1946) Selected International Committees	*ICOM-Conservation Committee (1967)* Selected Working Groups	*ICOMOS (1965)* Selected International Scientific Committees
Cleaning and Restoration of Paintings 1948–67 Scientific Museum Laboratories 1951–67		
	Mural Paintings, Mosaics & Rock Art (1959 as Mural Paintings)	Rock Art (CAR) 1980 Wall Paintings 1994
	Stone	Stone 1970
	Wet Organic & Related Materials 1984 (1961 as Waterlogged Wood)	Wood (IIWC) 1975
Museums of Archaeology and History and Historical Sites (now ICMAH) 1948		Archaeological Heritage Management (ICAHM) 1990
Documentation (CIDOC) 1950	Documentation 1975 (or before)	Documentation (1969) Architectural Photogrammetry (CIPA) 1969
Architecture and Museums Techniques (ICAMT) 1948		
Training of Personnel (ICTOP) 1968 (Administration and Personnel 1951)	Training in Conservation & Restoration 1969	Training (CIF) 1984
Glass	Glass and Ceramics	Stained Glass
Marketing and Public Relations 1976		Economics of Conservation 1988
Management (INTERCOM) 1989	Interim Working Group on Legal Issues 1999	Legal, Administrative & Financial Issues (ISCLFA) 1997
Museum Security (ICMS) 1974	Preventive Conservation 1993	Risk Preparedness 1997
Historic House Museums (DEMHIST) 1998		Historic Towns and Villages (CIVVIH) 1982 Historic Gardens 1973

(ICOMOS was, however, absent), which led to publication of a joint statement clarifying the status and aims of each organization.[27]

Since then, there have been many professional contacts, including representation on each other's governing councils. But only occasionally have there been joint meetings of the relevant committees/working groups of different organizations. For example, in 1991, ICCROM organized a workshop in Ferrara for the different international groups concerned with professional training in conservation – ICTOP (International Committee for Training of Personnel) of ICOM, CIF (Comité International pour la Formation) of ICOMOS, the ICOM-CC working group on training, and ICCROM's own Academic Advisory Board. In 1996, the Blue Shield was created to co-ordinate disaster preparedness and response policies between the four NGOs of ICOM, ICOMOS, IFLA (International Federation of Library Associations and Institutions), and ICA (International Council on Archives) (UNESCO and ICCROM have consultative status). But these joint activities have proven to be the exception rather than the rule. The overlapping of interests and activities already pointed out in the 1960s has, if anything, increased with the creation in recent years of new specialized groups.

Integrating the movable and immovable

This review of the development of the leading international organizations concerned with conservation shows that the movable:immovable distinction has been in evidence, but has not been perhaps a hard and fast one. Although used in the terminology of early UNESCO standard-setting texts, it did not lead to strong operational distinctions. Nowadays, it is common to associate ICOM with museum collections (the movable) and ICOMOS with the built heritage (the immovable). But, as we have seen, the history of their development is more complicated than that. ICOM was concerned with all collections and all forms of interpretation and education, including those in monuments, zoos, and parks; whereas ICOMOS was created for architects and engineers working on the technical restoration of buildings and sites. The Rome Centre, on the other hand, was founded for the conservation of all cultural property, with no reference to movable and immovable distinctions.

Now known as ICCROM, it keeps to this mandate. As mentioned earlier, ICCROM was founded with a mandate referring to the preservation of all cultural property. This was so, even though its creation was recommended by the UNESCO International Committee on Monuments, and even though it was the particular needs of monuments and sites that were felt to be under-served at the time. Fortunately, however, its given mandate was a broad one and so it has remained, covering the conservation of everything from the largest cultural landscape through to the individual postage stamp.

Nowadays, ICCROM is actively attempting through its programmes to cut across any demarcations that might exist within conservation, whether in perception or in reality. In late 2000, it defined a series of strategic directions that would guide the specific goals of ICCROM over the following four to six years. Number 4 of these strategic directions is to adopt policies and activities that integrate the conservation of movable and immovable cultural property.[28]

There are many ways to approach this goal. Two have already been implemented in the education area – one is to design programmes that emphasize issues that are common to all conservation, addressed to a mixed audience of movable and immovable specialists; another is to concentrate on materials, their properties, deterioration, and conservation, since these are common to movable and immovable objects.

An example of the first goal is the introduction of courses on the topics of Sharing Conservation Science and Sharing Conservation Decisions. The Decisions course focuses on how decisions are taken in conservation: Who are the players? What is the goal of the conservation project? What role does scientific analysis play? How do the art historians and architects provide their opinions? How do all the parties communicate? How are the decisions taken? Recently-completed or current projects are visited as case-studies, with the relevant players describing their roles in making decisions.

The other educational approach involves teaching that focuses on materials rather than the finished product. For example, the International Course on the Technology of Wood Conservation, organized every other year with Norwegian partners, is as relevant to objects conservators as to architects.

Despite these initiatives and the work of the very many other organizations that successfully ignore such categories, there is still some way to go before the movable:immovable distinction loses all meaning. It will doubtless remain entrenched in terminology and titles for some time to come. Nevertheless, the origin of the distinction in a particular legal system should not be allowed to inhibit an integrated approach to conservation of all cultural heritage expressions regardless of their mobility.

Biography
Nicholas Stanley-Price MA, DPhil
Dr Nicholas Stanley-Price has been Director-General of ICCROM in Rome since August 2000. He carried out archaeological research and administration in the Middle East for 12 years before specializing in conservation and professional education. He has held positions at ICCROM (1982–86), the Getty Conservation Institute (1987–95), and the Institute of Archaeology, University College London (1998–2000). He has edited various books, including *Conservation on Archaeological Excavations* (ICCROM, 1984; reprinted 1995), *Preventive Measures During Excavation and Site Protection* (ICCROM, 1986), and (with K. Talley and A. Melucco Vaccaro) *Historical and Philosophical Issues in Conservation of Cultural Heritage* (Getty Conservation Institute, 1996). He founded and edits the journal *Conservation and Management of Archaeological Sites*.

Acknowledgements
I am very grateful to Rachel Burch for extensive research assistance for this paper. We are indebted to many colleagues for information concerning international committees and working groups, including Isabelle Brajer, Patrick Boylan, Brian Egloff, David Grattan, Jukka Jokilehto, Star Meyer, Isabelle Pallot-Frossard, Andrew Powter, Herb Stovel, Hans-Christoph von Imhoff, and Werner von Trutzschler.

Notes
 1 Prott, L.V. and O'Keefe, P.J., '"Cultural heritage" or "Cultural property"'?, *International Journal of Cultural Property*, Vol. 1 No.2, 1992, pp. 307–20 (p. 312).
 2 O'Keefe, P.J. and Prott, L.V., *Law and the Cultural Heritage: Vol. 1 – Discovery and Excavation*, Professional Books, Abingdon (1984), p. 153.
 3 Prott and O'Keefe, *op. cit.*, pp. 307–20.
 4 O'Keefe and Prott, *op. cit.*, pp. 182–83.
 5 Baghli, S.A., Boylan, P. and Herreman, Y., *History of ICOM (1946–1996)*, ICOM, Paris (1998), p. 8.
 6 Boothroyd Brooks, H., *A Short History of IIC. Foundation and Development*, The International Institute for Conservation of Historic and Artistic Works, London (2000), pp. 3–4.
 7 International Museums Office, *Manual on the Conservation of Paintings*, International Institute of Intellectual Co-operation, Paris (1940) (French edition, 1939; English reprint, Archetype Books with ICOM, London, 1997).
 8 Iamandi, C., 'The Charters of Athens 1931 and 1933: Coincidence, Controversy and Convergence', *Conservation and Management of Archaeological Sites*, Vol. 2, No. 1, 1997, pp. 17–28.
 9 International Museums Office, *Manual on the Technique of Archaeological Excavations*, International Institute of Intellectual Co-operation, Paris (1940).
10 Baghli et al., *op. cit.*, pp. 15, 43.
11 Ibid., pp. 15–16.

12 ICOM Statutes 1960 (November 1961); see
   http://www.icom.museum/organization.html (August 2002).
13 Boothroyd Brooks, *op. cit.*, pp. 15–20.
14 Ibid., pp. 16, 23.
15 Ibid., pp. 21–22.
16 Ibid., p. 46; as defined in its statement of 1973 (see note 26 below): 'The Institute is concerned with the whole field of inanimate objects considered worthy of preservation, whether in museums or libraries or exposed externally, their structure, composition, deterioration and conservation'.
17 A list of the early meetings of the ICOM Commissions/Committees is given by Winter, J., 'List of Meetings', Supplement, ICOM Reports on Technical Studies and Conservation, *Art and Archaeology Technical Abstracts*, Vol. 14, No. 2 (Winter, 1977), pp. 378–79.
18 Pane, R., 'Some Considerations on the Meeting of Experts held at UNESCO House, 17–21 October 1949', in: *Monuments and Sites of History and Art and Archaeological Excavations. Problems of Today*, UNESCO (reprint, 1953; first published in *Museum*, Vol. 3, No. 1, 1950), pp. 49–89.
19 Lee, R.F., 'Report on the Findings of the Meeting of Experts', in: *Monuments and Sites of History and Art and Archaeological Excavations. Problems of Today*, UNESCO (reprint, 1953; first published in *Museum*, Vol. 3, No. 1, (1950), pp. 92–94.
20 *International Principles Governing Archaeological Excavations. Preliminary Report.* UNESCO/CUA/68, UNESCO, Paris (1955).
21 Daifuku, H., '"The Rome Centre" – Ten Years After', in: *The First Decade 1959–1969*, International Centre for the Study of the Preservation and the Restoration of Cultural Property, Rome (1969), pp. 11–18.
22 Jokilehto, J., 'The Context of the Venice Charter (1964)', *Conservation and Management of Archaeological Sites*, Vol 2, No 4, 1998, pp. 229–33.
23 Lemaire, R., 'Report of the President of ICOMOS, Raymond Lemaire 1975–1981', *Thirty Years of ICOMOS*, ICOMOS Scientific Journal, 1995, pp. 93–98; Jokilehto, J., 'The Context of the Venice Charter (1964)', *Conservation and Management of Archaeological Sites*, Vol. 2, No. 4, 1998, pp. 229–33.
24 ICOMOS Statutes 1978, Article 3; see
   http://www.international.icomos.org/e_statut.htm (August 2002).
25 Boothroyd Brooks, *op. cit.*, pp. 57–58.
26 *The First Decade 1959–1969, op. cit.*, pp. 25–26.
27 Anon., 'IIC News Supplement', *Studies in Conservation*, Vol. 18, No. 1, February 1973, pp. 1–5.
28 See the ICCROM web site, www.iccrom.org.

Beyond the Divide – Experiences from Scottish Conservation

CAROL E. BROWN

Abstract

Scotland's distinctive integrated approach to conservation stems from the work of a few key individuals in the field from the 1960s onwards, who applied a single standard across artefacts and buildings conservation. The inspiration for a single centre for conservation research, practical work, training, information, and advice spanning private and public sectors came from this tradition. The work of the Technical Conservation Research and Education Group (TCRE) of Historic Scotland and the Scottish Conservation Bureau are outlined in this paper.

The Bureau's Conservation Internship scheme has been transferred successfully from artefacts to the buildings sector in training building repair specialists as sole practitioners. Attempting to transfer conservation skills and knowledge into existing building crafts and trades is not easy. Current schemes for accrediting conservation professionals and practitioners in Scotland should aim to achieve a parallel approach rather than developing in isolation.

Case studies from the Bureau and TCRE illustrate the need for more integrated skills training and professional development across the building–objects 'divide'. The role of a Scottish Conservation Forum in Training and Education is promoted as a means of contact and a forum for finding common ground.

1974 – Scotland's Conservation Centre

From Scotland's point of view, I like to think that 'Conservation met Conservation' in Edinburgh in 1974. I see this as a formal meeting only, as I suspect that for decades this had been a close relationship. In particular, I visualize this as a meeting between two singular men, one an architect, James Simpson (Figure 1), and one a conservator, Robert (Rab) Snowden (Figure 2).

Figure 1 (left) James Simpson, founding partner
of Simpson and Brown Architects, Edinburgh.
Figure 2 (right) Rab Snowden at work in 1970.

In 1974, Simpson and Snowden together wrote a paper entitled 'A
Scottish Conservation Centre',[1] which was the starting point for a research
project cumulating in a working group report and a seminar at Hopetoun
House, West Lothian, in December 1976. The concept of a Conservation
Centre to fulfil all Scotland's conservation and crafts needs arose from a
move by the Council of Europe in the late 1960s to urge member govern-
ments to promote centres to co-ordinate the availability and training of
architects and craftsmen. The Simpson and Snowden paper indicates
that there was confidence in the idea that a Scottish Centre – especially
one related closely to an established centre of practical expertise – could
serve the wide variety of conservation and ancillary needs that appeared
to exist in Scotland at that time, and that it would fit very well into what
was identified as 'a developing regional network'.

Starting out in my post in Historic Scotland in 1992, I was keen to delve
into the history of the initiative that had set up the Scottish Conservation
Bureau back in 1980, and unearthed the paper and proposal in my first
week at work. As a conservator from a background in museums, I was
astonished by the scope of the content of contributions to this meeting
and the paper preceding it. The proposal for the setting up of a
Conservation Centre and the subsequent discussions were underlaid by

an unselfconscious assumption that Scotland's heritage encompassed buildings, objects, and landscapes. In addition, the need for the application of conservation science and preventive conservation input was acknowledged alongside the importance of redressing disappearing crafts and trades. The authors encompassed both the private and the public sectors within their paper, and treated the need for better conservation information as seriously as they did the shortage of building materials and traditional skills. It was a truly inclusive approach and, as such, well before its time. I realized that I was indeed in a different country.

The case for the Centre was founded on:

> ... an increased public interest in conservation generally and in historic buildings and works of art in particular; the severe shortage of conservation facilities for artefacts and works of art outside the Nationals; the decline of traditional craftsmanship and decreasing availability of materials; the shortage of professional and practical skills and of facilities for training architects, restorers, technicians, clerks of work and craftsmen and the lack of a central "clearing house" for relevant information.[2]

The Centre would be specifically concerned with:

> The conservation of decorative interior work: joinery, paintwork, plasterwork, wallpapers, tapestries and fabrics, furniture, pictures, prints and drawings. The conservation of buildings, survey and analysis, diagnosis of defects, traditional materials and methods of construction, techniques of repair and restoration and planned maintenance. The maintenance of traditional building crafts and material; stonemasonry, carpentry, slating, plumbing, tiling, harling, plastering, painting and ironwork.[3]

A comprehensive and all-encompassing list, you will agree.

Scotland's distinctive integrated approach to buildings and objects conservation

In Scotland, the different professions in conservation have always worked closely together; more so, I believe, than in other part of the British Isles.

One reason is simply due to the size of the country and its geography and communications. A tradesman can be self-sufficient with a range of useful building skills within a remote area, but in conservation you need also to be able to make good contacts if you want work outside your area; and there still is an excellent close network that exists in the heritage field north of the border. The success of the Scottish Society for Conservation and Restoration (SSCR) over the years attests to this – it has a geographical remit, but has always encompassed and been run by a broad cross-section of architects and conservators. Again, SCCR was founded early in 1977 and two of the founder committee members were Simpson and Snowden.

The effect of this cross-sector remit can be tracked over the years; I have seen a steady reduction in SSCR membership over the last few years as the organization has chosen to align itself more with objects conservation to keep in step with accreditation moves in the rest of the British Isles and in Europe.

It is also clear that the late 1960s and early 1970s was the era when those concerned with the heritage came to realize the full consequences of the changes that had occurred in the organization and methods of the construction industry since the Second World War. The phasing out of traditional materials and skills was also exacerbated by the levels of inflation prevalent at that time. These losses were probably felt more keenly in Scotland where the basic traditional building materials are stone and slate.

I suspect another reason for a more integrated approach in Scotland was that the various strands of professional and practitioner were bound by a common crafts background with a dedication to high standards. The nature of the indigenous building materials used in Scotland left a legacy of excellent stone masonry, carving and slating skills, now in decline, but subjects that are still under-represented per capita south of the border. In addition, the influx of academic science-trained conservators into major institutions (if not the private sector) in Scotland had been steady since the 1950s, but generally slower than in other parts of the United Kingdom. Scotland previously built on a tradition of training and apprenticing restorers in disciplines related to high-standard trade backgrounds – particularly paintings restoration, signwriting, and the production of specialist paint finishes and decorative techniques.

The area of conservation and restoration of interior schemes and structural decoration is, in fact, another useful link back into our theme of 'Conservation meeting Conservation'. Scotland's expertise in this area was driven in particular by the wonderful heritage of seventeenth-century painted beam and board ceilings and other structural paintings in buildings throughout the country; particularly those in the care of the National Trust for Scotland (NTS) (Figure 3).

This leads us to the other uniting force for conservation and conservation in Scotland: Stenhouse Conservation Centre at Stenhouse Mansion in Edinburgh. Set up by the NTS in 1967 and taken over by the then Department of the Environment in 1975, the Centre focused on the conservation of paintings and stonework relating to the properties and monuments in the Government's care as well as keeping a remit for the Trust's properties. To quote the Simpson and Snowden paper:

> It is to Scotland's great advantage that the Stenhouse Conservation Centre was set up by the National Trust for Scotland to work in the area which traditionally lay between the realms of architect and the picture restorer – that of structural decoration.[4]

Conservation of interior schemes is necessarily a place where buildings and objects conservation do meet; in 1967, the restoration of the Mansion itself and the creation of the Centre were a meeting point for architects, craftsmen, and restorers alike and formed the inspiration for the 'new' Centre proposed in the 1974 paper.

And the culmination of these proposals? It is interesting to see how these deliberations worked out in practice over the years. The scheme was indeed far-reaching – too much so, perhaps, and impractical in its idealism and scope . 'A young person's initiative', as James Simpson has said to me. Funding could not be found, but parts of the proposal continued to be nurtured by supporters such as Lord Bute, Barbara Whatmore from the Radcliffe Trust, and Judith Scott from the Council for the Care of Churches at the time. The feeling then was that the networking

Figure 3 Fellow Fred O'Connor working on a beam and board ceiling at Huntingtower.

element was the most important factor – the means of bringing together the client with the contractor; the answer did not have to be in the form of an institution.

With backup from the Crafts Council of Great Britain, the Conservation Bureau was born within the Crafts Division of the Scottish Development Agency in 1980. Dick Reid, the stone and wood carver from York, was the link between the Crafts Council and the equivalent body in Scotland and, as a strong proponent of the private sector, he helped give the Bureau its distinctive overview of the wider Scottish crafts building and artefacts sectors.

Ten years later, with an admirable symmetry, the Bureau was 'captured' and brought into the Stenhouse Conservation Centre (quite literally) by the foresight of Ingval Maxwell, and a year later both Centre and Bureau were rolled into Historic Scotland's new Technical Conservation Research and Education Division, under Maxwell's Directorship.

How Interns and Fellows have brought conservation together

The history of the Conservation Bureau's internship and fellowship schemes is one area that can throw more light on this subject.

The internship scheme began in 1987, when the Bureau was in its former home in the Crafts Division of the Scottish Development Agency. Two bursary schemes were started at the Burrell Collection in Glasgow to provide a two-year in-house training programme to address the shortage of private-sector conservators in ceramics and furniture. When Historic Scotland took over the Bureau in 1992, we began to extend the scheme – seeing it as an excellent direct way of increasing the fund of skills and conservation knowledge in Scotland and targeting subjects where Bureau enquiry data showed there was a shortfall of practitioners to answer a demand – looking solely at the portable heritage (Figure 4). The Bureau's job was to seek placements for newly-trained conservators to work alongside experienced practitioners for a 12-month period – using the public sector at first, but later branching out into the private sector. Interns were encouraged, as far as possible, to stay on in Scotland and over the years (with five or six interns per year) we have had roughly a 75 per cent success rate. Since 1996, we have had a contractor in place to manage the scheme, to provide support to interns and supervisors, and to ensure accountability for Historic Scotland.

At the same time that the scheme was beginning to be developed within the Bureau, we also had approaches from building contractors working in the areas of traditional repair, particularly limework – a relatively new

Figure 4 *Textile conservation intern, Sarah Foskett, with supervisor Helen Hughes at Glasgow's Burrell Collection.*

and lively area. They had heard of the internships; would we run one in the building field? Our reaction to this at the time was mixed: we knew how to run an internship and the course of the year was predictable; the candidate was guaranteed to be trained, if not experienced, to a certain level. Moreover, we had a good idea about what the outcome would be – someone more likely to be able to set up in business on their own or be eligible for employment in the museums or other heritage fields in Scotland.

With buildings conservation – yes, it was clear that there were just as many shortages of skills and over a much broader area of disciplines; we knew there were also useful contractors out there willing to train them. Our major worries about transferring the scheme to the buildings arena were to do with input and output: what kind of person were we looking to recruit and what kind of contractor were we hoping to put out into the world? Where would they work and what was the demand?

Looking back, I think we made things difficult for ourselves by beginning the scheme with 'traditional building repair' in mind, rather than the individual construction trades or crafts. Traditional building repair is not easily classified as an area of the construction industry; there are no formal entry routes, training, and qualifications, and one needs a wide range of skills and a knowledge of a lot of different materials and building types. It is an interesting niche that is important to Scotland, in particular, due to the nature of the building stock and materials; also because being 'jack of all trades' is a necessity when you live and work in remote areas. But what was the demand for this work? There was not sufficient data.

Another factor in setting up a new scheme was the length of time needed. With interns, we knew that a year was adequate to give valuable practical experience to students who already had conservation academic training, but little hands-on experience. In the building conservation area this was another issue. We felt that to 'convert' someone from perhaps a trade or other craft background to conservation thinking would not necessarily take long, but we would need to apportion two years to cover the breadth and depth of skills training needed, as well as time to accommodate the seasonal nature of this kind of work. Once we decided to trial this scheme, we felt we needed another name too, to distinguish this from the internship (and looking back I do wonder why, considering the theme of this conference) – and so the Historic Scotland Building Conservation Fellowship was born.

In this area we have been successful over six years in recruiting seven fellows – four women and three men from various trades, such as stone masonry and plastering, or backgrounds such as practical environmental work, objects conservation, and practical art. They have been based as 'working pupils' at the Scottish Lime Centre (SLC) at Charlestown Workshops in Fife. The SLC provided a base and initial training, and went on to organize placements throughout Scotland, tailored to the pupil's background and requirements. Fellows also attended the conservation materials and philosophy lectures at the Edinburgh College of Art's Scottish Centre for Conservation Studies (Figure 5).

All seven fellows have gone on to conservation work in Scotland (alas, one has subsequently left to work in England, but he promises to return) – one becoming a surveyor, two becoming stone conservators, others filling the building repair niche. And that has brought us to the question – what have we created? Certainly a handful of sole contractor building conservators who are keeping alive traditional skills while working to a conservation ethic. They are doing a valuable job, often providing the skills and experience to take on work for the major conservation agencies dealing with historic buildings and monuments.

Nevertheless, we are aware that there is a wider area that remains to be tackled – how do we 'infiltrate' and influence the wider construction industry in Scotland? In Historic Scotland, we are tasked to promote high

Figure 5 Fellow Becky
Little with Tim Meek at
Brodie Castle.

standards in conservation and the use of traditional skills and materials;
within the Technical Conservation Research and Education (TCRE)
Group we are using our research and publications, our links to training
courses, and materials initiatives such as the creation of the Scottish Stone
Liaison Group[5] to promote the use of Scottish natural stone. How far
are the fellowships contributing to this work?

It may be interesting to note, therefore, that our latest plans for the
scheme involve returning to something closer to the internship model
and a recognition that the individual trade or craft area should be
paramount, rather than a wide-ranging syllabus. We plan to place each
apprentice-level student directly with a practitioner in the individual craft
and trade area (we are starting off with plastering and architectural
joinery) for a two-year period. The conservation content of the
programme will come from modules spent with the Scottish Lime Centre
Trust, the Centre for Scottish Conservation Studies, or Historic
Scotland's own conservators, concentrated in the winter months. With
this programme, the link with the supervisor will be paramount, with
in-house supervision and administration within TCRE to ensure 'pastoral
care' for the fellow and their supervisor and accountability to Historic

Scotland. In this way, the portable heritage conservation model is being used to inform the buildings model, after all. We had assumed a different approach would be necessary, but, in reality, the needs were identical in both sectors.

Accreditation – parallel lives

Conservators
I am fortunate to have been involved with the development of the portable heritage conservators' accreditation scheme – National Council for Conservation-Restoration's (NCC-R) Accreditation of Conservator-Restorers (ACR) system,[6] since its furthest beginnings in 1986 and had been involved with the Museums Training Institute, later Cultural Heritage National Training Organisation (CHNTO), working group on standards in conservation of the portable heritage.[7] By 1996, I was involved with the United Kingdom Institute for Conservation (UKIC) Accreditation Working Group and, later, as observer for Historic Scotland with the Joint Accreditation Group. We set up focus groups and training in Scotland throughout 1999, and in 2001 in conjunction with SSCR to test the ACR scheme, to train assessors, and attract potential candidates. In 2002, we have a system in place that is well thought out and based on the concept of the 'reflective practitioner'.

The conservation profession's only problems with this scheme are marketing and reconciling the various parallel uses of the system with others run by the various discipline-specific bodies within our own very small field. The take-up for the full scheme has been slow, despite the clever initiative of running a 'fast-track' system in advance to gain some critical mass. Though widely promoted in 'conservation' literature, my personal experience is that Professional Accreditation of Conservator-Restorers (PACR) in general, remained a development in camera and remote from other professions – especially those outside the museum world. From my rather ambiguous status as émigré from the artefacts side of the conservation-conservation 'divide', I have been amazed over the years by conservators' lack of assurance in not looking outside their own field and at the same time incensed by the disregard many architects have for what we call the 'real' conservation world.

The lack of cohesion and marketing of NCC-R's ACR system is now being tackled head-on. In 2002, steps were taken to bring together the common elements of each scheme in a simple framework. The obvious outcome of this has been to provide a single understandable and marketable system available to conservators in all disciplines.

Building professionals
And this is a real grey area between conservation and conservation – or I can now say it has been until very recently. Over roughly the same period, in the architectural and building conservation world, several schemes have been set up to recognize conservation competencies of professionals in the professions. These include schemes run by the Architecture and Surveying Institute (ASI), Architects Accredited in Building Conservation (AABC), Royal Incorporation of Architects in Scotland (RIAS), and the Royal Institution of Chartered Surveyors (RICS). The Institute of Civil Engineers (ICE) and the Institute of Structural Engineers (ISE) are currently working towards a scheme of their own. There is also the Institute for Historic Building Conservation Scheme, set up some years ago. The major difference between these groups and the objects conservators is that, in buildings, individuals have largely been accredited already within their own profession – conservation being seen as an optional add-on.

While these professionals acknowledge that their new accreditation schemes have served a useful purpose, there is also widespread concern that the establishment of a coherent, sustainable, and common system of professional accreditation in the conservation element is still some way off. The number of conservation-accredited practitioners is acknowledged to be disappointingly low and those involved are worried by how assessment systems might work and how to make clearer links between accreditation systems and training. Peer-group review has been the preferred method of assessment for these schemes, requiring individuals to submit a portfolio of evidence from five projects upon which their abilities are judged. This system has proved problematic and difficult to defend from accusations of nepotism or elitism. It all sounds very familiar.

In essence, the building professions working in conservation are going through the same process as conservators, a few years behind perhaps, but this is not surprising bearing in mind the substantially broader field of building-related professions, the greater numbers potentially involved, and the financial influence these professions command.

Constructing the bridge
This is where the portable heritage conservators can learn and can benefit, perhaps. In the specifiers' field, there is a good deal more leverage and incentive available from a financial and political point of view. Both English Heritage and Historic Scotland have intimated the necessity for all lead professionals (in the first instance) working on grant-aided cases

to be accredited over the next couple of years; the Heritage Lottery Fund is expected to go down this route as well. The accreditors in the building field also recognize that there is a consumer demand for a conservation hallmark – especially among property owners – for a means of identifying professional advisors with conservation skills and experience. A need far wider than the relatively small amount of work that is grant-aided. It needs to be noted that expenditure on maintenance and repair of the building stock now accounts for more than half the British construction industry's total output – amounting to some £28 billion in annual turnover.

The move towards trying to get these two areas to 'meet' began with Ingval Maxwell in Historic Scotland about ten years ago. The current stage of this initiative, from 2001, has involved on-going discussions with current accrediting bodies to bring the various building schemes together into a common framework, setting up an Accreditation Working Group. This has fostered the production of two key documents. One, a Model Accreditation System, based on the key parts of the frameworks of existing building-related schemes, with the aim of creating a flexible model system that could be applied to any of the professions concerned with building conservation. The second document was one commissioned by Historic Scotland from Heriot-Watt University and Edinburgh College of Art (SCCS) aimed at devising a framework upon which individuals could be guided in the preparation of their evidence. This used the ICOMOS training and education guidelines of 1993 as a basis for a Continuing Professional Development (CPD) framework,[8] to be used as a means of supporting practitioners in compiling their portfolios and identifying areas where their experience could be extended.

Another key factor for the buildings side was the publication by English Heritage in 2000 of *Power of Place*,[9] which recommended the setting up of a national conservation training forum to bring together all training and qualification initiatives.

The Buildings Accreditation Working Group had been working for over a year before the parallel work of NCC-R was noted and found to be relevant – this happened earlier in 2002. Since that time, I am pleased to report that the connections between the two sectors of conservation have been recognized fully, with the invitation of a NCC-R representative to the meeting of the group in August 2002. Driven by a strong desire not to re-invent the wheel, the buildings sector is looking seriously towards the NCC-R scheme as a model for a way forward.

A way forward – learning each other's languages

In conclusion, therefore, I feel it is more useful to talk about the points of connection rather than differences.

The three areas I have touched upon are all examples of meeting places where conservation has learnt from or been inspired by the other 'side'. The theme of this conference is surely about opportunities – in my experience there are two groups of people out there who are working towards the same ends – often working on the same project and communicating very well in individual work situations. It is surely the separate development and the history of each profession that has led the two strands apart via separate routes for training, accreditation, and research, and caused us to speak different languages.

It is surely only a linguistic difference that we have to overcome to make this cultural reconnection, and I hope I have shown by our Scottish experiences that the idea can become reality – in time!

Biography

Carol E. Brown BA, Dip Arch Cons, ACR

Carol Brown trained in archaeology and later in archaeological conservation at Durham University before working as a conservator at museums in Salisbury, Cardiff, and Bristol. She worked in the south west, specializing in training and advisory work with curators and volunteers, before moving to Scotland in 1992. Carol Brown currently heads a team supporting conservation of buildings and objects across Scotland and providing a conservation advice service. During the 1990s, she served on the MTI/MGC Training Panel devising conservation standards and S/NVQ qualifications in conservation and also the UKIC Accreditation Working Group relating to the UKIC's 'fast-track' accreditation system.

Notes

1 Snowden, R. and Simpson, J., 'Scottish Conservation Centre Report', *Hopetoun House Seminar*, 14 December 1976, unpublished discussion paper, Edinburgh (1977).
2 Ibid.
3 Ibid.
4 Ibid.
5 Maxwell, I. and Ross, N., 'Foreword to Conference Proceedings', *Historic Scotland Traditional Building Materials Conference*, Historic Scotland, Edinburgh (1997).
6 *Professional Accreditation of Conservator-Restorers: Introduction*, National Council for Conservation-Restoration, London (1999).
7 Museum Training Institute, *Standards of Occupational Competence for the Museums, Galleries and Heritage Sector*, MTI, Bradford (1996).
8 ICOMOS, *Guidelines for Education and Training in the Conservation of Monuments, Ensembles and Sites*, ICOMOS (1993).
9 English Heritage. *Power of Place: The Future of the Historic Environment*, English Heritage, London (2000).

'Places' and 'Stuff': Is it Only the Language of Conservation that is Changing?

MAY CASSAR

Abstract

Taking the three thematic areas of sustainability – social, economic, and environmental – as the starting point, the author considers the wider context within which conservation of the material heritage is practised. In this, it is argued that similar influences and pressures affect society, the natural environment, and the material heritage, yet nature conservation has both 'natural' and 'man-made' protection. Nature can renew itself if well managed, while material heritage cannot, and organizations supporting nature conservation have managed to raise public and political awareness in a way that conservation of the material heritage has not. The author argues the merits of conservation of the material heritage as being the management of a non-renewable resource, and that aligning material conservation with the ethical principles of sustainability will provide the societal context for wider recognition. It is also argued that achieving a balance between conservation and access, as a cornerstone of heritage sustainability, will become a more realistic goal if conservation practitioners identify more closely with society's interests. One way in which this can be done is by involving communities in conservation decisions. The paper uses examples of work undertaken by University College London's Centre for Sustainable Heritage to illustrate certain points.

Introduction

The evolution of a language often reflects wider changes in society, industry, or a profession. Conservation is no exception.

Growing professional maturity and wisdom are evidenced by the wide acceptance today of the need to consider the whole object rather than single elements; of risk assessments that consider acceptable levels of damage rather than analysing every damage incident; acknowledging that many historic materials are robust and therefore can be made freely accessible; and that sometimes we may overestimate the fragility of objects. Additionally, that we need more critical assessment of the balance between preservation and access rather than preservation at all costs; recognition of the value of working in interdisciplinary teams and the importance of context and community for long-term sustainability of the heritage.

Conservation is undoubtedly a shared responsibility. It is as much about people and how they interact as it is about buildings and collections. It involves applying different disciplines to a common cause and it may require difficult compromises to achieve greater, better, and longer-lasting results. These developments have made conservation a more complex process, but also a much deeper and richer experience. It is also making it more responsive to a changing society that is increasingly in tune with its environment.

A changing context

Our world perspective has changed in recent years from one of universality, a single approach, and the belief that reality can be modelled and understood through a discrete set of parameters, to one in which we recognize the complexity and dynamism of the world we inhabit and its variety of species and cultures. We have a global (i.e. world-focused), rather than a universal perspective, in which diversity is a key issue, whether it is cultural or ecological, and where sustainability is a social, economic, and political force to be reckoned with.

These ideas are also echoed in the world-wide nature of conservation, how it transcends geographical and political boundaries, and how exciting it is to observe those with a common cause develop a common approach to planning and education and finding solutions to long-standing problems. This perspective accepts that it is impossible to control everything. Standards are being challenged by methodologies and processes that are transparent and consistent. Universal solutions, it seems, are not the answer; deterministic approaches and an eagerness for standardization oversimplify reality.

There are other challenges to be faced: changing societal needs demand that we define conservation in terms of 'quality of life' of citizens and communities; that while heritage preservation is valued by society, this is primarily in order to sustain a sphere of public interest and public good. Through all these societal shifts and changes, the philosophical and ethical principles upon which all conservation activity is constructed still stand, though modified and reinterpreted.

The amendments to the *Burra Charter* in 1999 overtly recognized that heritage value and significance are embodied in the uses, meanings, and associations of a place, in addition to the physical fabric of a place or structure.[1] This represents a significant shift towards integrating the tangible and intangible heritage.

So how are these paradigm shifts in conservation thinking influencing conservation decision-making? Here the now familiar concept of cultural capital is a useful one. Inherited from past generations, cultural assets may be used, but not consumed, because they are held in trust for future generations. Like natural resources, if these assets are not maintained, they will decay and lose their future value. Re-defining conservation in socio-economic terms sets it clearly within a context of sustainability and provides an engaging contemporary interpretation for conservation.

Development of a sustainability agenda

Differentiating sustainable development and sustainability

There are, however, conceptual and practical difficulties in trying to understand the principles and application of sustainable development because the concept is, at the same time, complex and vague, amorphous yet populist. Its definition changes depending on whoever is defining it and at what level – internationally, nationally, regionally, or locally. Nevertheless, it is clear that 'development' can imply growth, and growth can lead to pressure on the world's resources to meet a growing need.

Sustainability, on the other hand, is sometimes defined too simplistically; it is used interchangeably with the word 'survival', saying 'can it be sustained?', when we mean 'can we go on?' or 'can we afford it?'

Sustainable development is a process – it is not an end in itself. The central aim of sustainability is to achieve an acceptable quality of life for the world's population, combined with economic growth of communities, without depleting or damaging the natural resources needed to sustain future generations. The link between sustainable development and sustainability is one of costs and benefits. Applied to the historic environment, one might ask 'What are the costs to the heritage, the present environment, and the future of the planet of unsustainable practices; and what are the benefits of sustainability?'

Growth in international interest in sustainability

The first tentative steps towards defining sustainability in terms of its social, economic, and environmental elements were taken at the United

Nations' Conference on the Human Environment in Stockholm during 1972. In 1987, the United Nations' Commission of Environment and Development produced *Our Common Future*,[2] more commonly known as *The Brundtland Report*, whose most memorable quote is the definition of sustainable development – 'meeting the needs of the present without compromising the ability of future generations to meet their own needs'. A re-definition in conservation terms could be 'the process of maintaining and managing with others the use of cultural heritage for the benefits of today's communities and future generations'.

United Kingdom Government promotes sustainability

Following the publication of its white paper on the environment, *This Common Inheritance*[3] in 1990, the United Kingdom Government published *A Strategy for Sustainable Development for the United Kingdom* in 1999.[4] The key tenets of this strategy are social progress that recognizes the needs of everyone; effective protection of the environment; prudent use of natural resources; and maintenance of high and stable levels of economic growth and employment.

The Government also produced a series of headline indicators[5] to monitor and report on progress towards sustainable development covering social progress, economic growth, and environmental protection, including people's everyday concerns – including health, jobs, crime, air quality, traffic, housing, educational achievement, wildlife, and land use.

In 2000, the Government passed *Amendments to the Building Regulations*, set out in 13 parts (i.e. Parts A to N, omitting I) to take on board more explicitly its sustainability agenda. For the first time the Regulations include buildings of historic character and significance. Part L, dealing with conservation of fuel and power, makes a direct link between global environmental protection, energy conservation, and the kind of measures that are acceptable to improve the internal environment of buildings, including historic buildings; for example, better insulation and a reduction in air infiltration before the use of energy-hungry systems. To support the implementation of Part L within the historic building sector, English Heritage worked with University College London's (UCL) The Bartlett and the Centre for Sustainable Heritage to produce guidance for local authority building control officers on the application of the revised Building Regulations to the historic environment.[6]

Linking heritage conservation and global sustainability

This is evidence of growing links between global sustainability and heritage conservation that can be further developed as long as we have the wherewithal to do so. Heritage conservation is inherently environmentally sustainable: the re-use of old buildings; the recycling of materials; avoiding waste; our awareness of the fragility of the air, land, and water; and that energy is a finite resource. Yet there is much that still needs to be done to reduce the environmental double standards that deliver heritage protection with one hand and environmental destruction with the other. When we specify energy-intensive systems, such as air-conditioning to control the environment within museum buildings for example, we contribute to the rise in carbon dioxide emissions leading to global warming.

Even today, an integrated approach to the management of the built environment (which does away with independent design and operation of the fabric and services, promotes the creation of one mutually dependent system, and recognizes that tight environmental control is not a foregone conclusion and that the use of older buildings should aim for a 'long life–loose fit' approach) is still the exception rather than the rule. The historic environment has the opportunity to lead the whole building sector as an exemplar of sustainability, by highlighting where historic buildings, contents, and occupants have managed to exist side-by-side in a sustainable way.

It is more difficult to assess the impact of the historic environment against the Government's social and economic indicators for measuring the country's progress towards sustainability: health, jobs, crime, traffic, housing, educational achievement, and economic prosperity. These indicators have been identified as of particular interest to the general public, and it behoves us to endeavour to bridge the language and perception gap between the heritage sector and other sectors – such as business, commerce, and industry – that have embraced the whole sustainability agenda as one, and that will not only help the planet, but also foster economic competitiveness.

Developing the sustainability credentials and limits of the historic environment

In 2001, a self-selected group of heritage bodies including English Heritage, The National Trust, the Heritage Lottery Fund, and the UCL Centre for Sustainable Heritage (and with the support of the Environment Agency, Countryside Agency and English Nature) began meeting to discuss the possibility of developing a 'Heritage Sector Sustainability Strategy'. Impetus was provided by the DTI/DEFRA's (Department of Trade and Industry/ Department of the Environment, Food and Rural Affairs) *Pioneer Group Programme for Sustainability*, an initiative created to encourage business and industry to adopt more sustainable practices.[7] A number of heritage organizations had already begun to develop their own approach to sustainability – for some, this has grown out of preservation, conservation, and/or environment activities into a wider ethical approach, whilst for others it has involved identifying specific activities such as archaeological site management, tourism, marketing, and architectural conservation that have similarities to business practices and which could usefully adopt a sustainable strategy.

The challenge for this group has been to explore how heritage organizations could benefit from an exchange of views with other sectors and how the discussions might help with the development of a sectoral sustainability strategy. At a series of meetings, the group assessed the relevance to the heritage sector of the sustainability challenges being faced by the business world, using language common to business and industry. One thing to note however is the difficulty in defining the limits of our responsibilities. Clearly, we already identify with recycled materials and re-use of buildings, but how much common ground do we have with other related sectors, such as nature conservation or the re-use of brownfield sites? How do we perform on issues relating to land or waste management or supply chain integration? As a sector we are dependent on suppliers of materials and services. What of *their* sustainability credentials? Where do we draw a line on *our* responsibilities? These questions are left hanging, but they are ones that we will all need to address before long, and some are doing so already. A personal view is that there is great potential for us to do much better than we have to date.

English Heritage's *Power of Place*[8] and the Government's response, *A Force for Our Future*,[9] address some of these issues. They state explicitly that conservation is not an end in itself, but about passing on heritage of maximum significance to present and future generations; that changes in collections, historic buildings, and archaeological sites affect their significance, authenticity, and value and that sustainability of the historic environment is about managing change so that as much as possible of the significance of a collection or a building is retained.

If we define sustainability as 'the reduction of environmental impact by not consuming non-renewable resources', we find that this also applies to the physical historic environment, which may be described as a non-renewable resource. Continued and new use of old buildings not only extends their productive life, but also conserves the embodied energy, human labour, and skills that went into creating them originally, and by preserving the spirit and memory of a place, they perform a crucial community function. If, on the other hand, extensive physical adaptation is required for the use of old buildings to continue, then an unacceptable level of loss in conservation terms may occur. There is a fine line to negotiate. Will there come a point when so much has changed physically that much significance and authenticity is lost? Or can significance, if not authenticity, be recovered in some way? The Buddha at Bamiyan may illustrate this point.

The Buddha at Bamiyan

Early in 2001, the Taliban carried out the deliberate destruction of the two giant stone statues of Buddha at Bamiyan in Afghanistan. These statues were carved some 1,500 years ago when the country was a key link in the Silk Road trade route. The world was powerless to stop their destruction. The Taliban intended to obliterate all reference to Afghanistan's pre-Islamic heritage and, by destroying the statues, their aim was to destroy Afghanistan's Buddhist past. It was recognized that the statues had significance beyond the aesthetic quality of their carving and the beauty of their setting – they had a religious and political significance. Their destruction appeared final.

Then, in January 2002, BBC Radio 4 announced that there were plans for the possible re-construction of the Bamiyan Buddhas.[10] Paul Bucherer, Head of the Afghanistan Museum in Basel (Switzerland) and one of the world's foremost experts on the Bamiyan Buddhas, returned from a UNESCO mission to Afghanistan and stated that, in meetings with government officials and talking to ordinary people in the markets, he had found widespread support for the idea of re-construction. The reasons are that the physical presence of the statues will have a symbolic significance, giving the Afghan people renewed meaning to their national identity, and, on a practical level, help to restore the local economy that benefited from the statues as tourist attractions in the years before the war.

Conservation and change

So, should the concept of conservation as a process of managing change include extensive material re-construction if it contributes to the maintenance of social cohesion? Is this an acceptable extension of the definition of conservation or does it go beyond the limits we are prepared to accept? It might be acceptable for example, that when the scale and complexity of the decision requires, communities most strongly affected by any physical changes to the historic environment will have the final say on what and how much is done in their name by experts.

These are the philosophical and ethical issues that provide the context for the research and teaching at the UCL Centre for Sustainable Heritage: how we find the right balance between physical preservation, continuity of use, and the meaning of heritage. We try to consider meaning, significance, context, and relevance of cultural heritage together with materials, condition, and aesthetics in our work and, in so doing, we try to overcome some of the traditional distinctions and barriers to collaboration.

Although the three key elements of sustainability are social, economic, and environmental, the basis of everything is still the environment. It provides the context for all social and economic activities, which are constrained by environmental limits. This is why our research focuses on the environmental sustainability of the historic environment, with the range of projects grouped under three headings – past climate, present climate, and future climate.

Past climate

We are working in partnership with English Heritage, The National Trust, and Historic Royal Palaces to profile the past environment within historic houses in an attempt to predict future internal conditions. This work is investigating the natural ageing of objects that have spent their lives in historic buildings in order to find links between object and environment over long periods of time. This will help us understand and predict the effects of environmental risks to historic objects. The work is tying together three separate strands of investigation:

1. Changes in the objects themselves.
2. Characteristics of the buildings in which the objects are housed.
3. Internal environment to which the objects have been exposed over their lives.

By also using archives of proxy data (i.e. data collected for other purposes, but interpreted in new ways), we are aiming to create a picture of internal conditions in selected houses. This includes inferring the indoor temperature and humidity in a house from historic fuel bills (such as coal and wood used at Kew Palace), and external conditions and documentation of commissioned maintenance and repair work held in account books (such as the frequency of internal repainting and re-plastering).

Present climate

We are also the co-ordinators of a European Commission research project to investigate how building design and materials determine indoor pollutant concentrations. 'Innovative Modelling of Museum Pollution and Conservation Thresholds' (IMPACT) is researching how the internal surface materials in a building interact with air pollutants through laboratory-based experiments, field measurements, and computer modelling. Early findings suggest that the role of building design and materials in controlling indoor levels of air pollutants has been underestimated, so the project is examining ways to enhance this passive pollution control aspect.

The main output of the project will be a publicly accessible website providing information (including case studies) for museums, galleries, and archives on the effects of air pollutants on heritage materials and how they can be controlled. At the heart of the website will be a computer model developed by the project to predict the concentrations of air pollutants inside buildings, which can take into account the building design, ventilation, and interior materials and finishes. Essential to its development is experimental work on deposition velocities of sulphur dioxide, nitrogen dioxide, and ozone on interior surface materials, as well as trials of new materials as passive pollutants absorbers in the museum environment, and field testing of the model itself.

Future climate

The world's climate is changing as a result of past and present human activity. In the face of complex problems and long-time scales, many are ducking the issue so that few detailed impact and adaptation studies have been carried out. Climate scientists expect that significant patterns of change will be established in about 50 years – for many in government, in business, and in society in general, this sounds too far away to be a present worry.

The heritage sector on the other hand is used to thinking in terms of centuries rather than years. Climate changes present problems that must be dealt with now when planning the future of the historic environment.

For this reason English Heritage commissioned a scoping study to assess the impact of climate change and the adaptation of the historic environment that will be required as a result. We are carrying out a year-long study, which includes engaging heritage managers in discussions about the climate changes predicted for the United Kingdom over the coming century and what these changes will mean for future planning: from detailed impacts, such as necessary changes to drains and rainwater downpipes, to new management and maintenance strategies and funding priorities. The report for English Heritage and scheduled for completion in April 2003,[11] forms the basis for further research defining problems and strategies for adaptation of the historic environment to climate change. Over the next three years, in a research project funded by EPSRC (Engineering and Physical Sciences Research Concil), the problem of drying out of historic buildings will be investigated. It will also serve as a model for adaptation studies for the wider building sector and planning community.

Conclusions

The conservation of the historic environment and the principles of global sustainability have much in common. This commonality provides an opportunity to demonstrate the relevance of the historic environment to society – it touches everyone's life. And it is not just the language of conservation that is changing; the world within which conservation functions has never been more receptive to saving the planet - conservation is on everyone's mind. We need to seize the chance these opportunities offer to move conservation of the historic environment up the list of everyone's priorities.

Biography
May Cassar MSc, FIIC, AMUKIC, FRSA
May Cassar is Director of University College London's Centre for Sustainable Heritage, working on the sustainable use of historic buildings, collections, and sites through research, teaching, advice, and consultancy. Her personal research interests include heritage sustainability, climate change and its impact on the historic environment, and predicting damage to historic collections. She was formerly Environmental Adviser at Resource: The Council for Museums, Archives and Libraries and the Museums & Galleries Commission, and is author and editor of six books relating to the interdisciplinary field of preventive conservation.

Notes

1 Australia ICOMOS, *The Australia ICOMOS Charter for the Conservation of Places of Cultural Significance* (Burra Charter), Australia ICOMOS (1999).
2 United Nations' Commission of Environment and Development, *Our Common Future: The Report of the Commission on Environment and Development* (Brundtland Report), Oxford University Press, Oxford (1987).
3 Department of the Environment, *This Common Inheritance: A Summary of the White Paper on the Environment*, HMSO, London (1990).
4 Department of the Environment, Transport and the Regions, *A Better Quality of Life. A Strategy for Sustainable Development for the United Kingdom*, DETR, London (1999).
5 Indicators for a strategy for sustainable development in the United Kingdom can be found on the following website:
http://www.sustainable-development.gov.uk/indicators/index.htm
6 The guidance can be found on the English Heritage website:
http://www.english-heritage.org.uk
7 Information can be found on the following website:
http://www.societyandbusiness.gov.uk/government/activities/detr/strategies.htm
8 English Heritage, *Power of Place: The Future of the Historic Environment*, English Heritage, London (2000).
9 Department of Culture, Media and Sport, *The Historic Environment: A Force for Our Future*, DCMS (Architecture and Historic Environment Division), London (2001).
10 Information can be found on the following website:
http://news.bbc.co.uk/1/low/world/south_asia/1790674.stm
11 The research results will not be buried in reports that will not see the light of day. We aim to make research results accessible on our website at www.ucl.ac.uk/sustainableheritage or through client publications.
The results will also be integrated into our teaching programme, short-course programme, and the new (from October 2003) MSc in Sustainable Heritage.

Working Buildings: The Effect of Building Use on the Conservation of Wall Paintings and Polychrome Surfaces

TOBIT CURTEIS

Abstract

Although our medieval churches and cathedrals are primarily working build- ings, their ancient fabric is highly vulnerable to deterioration, as are the historic artefacts contained within them. Often, the requirements of those using historic buildings are significantly different from those whose role it is to maintain and conserve them.

The expectations of modern congregations and visitors are very different from those of only a generation ago, and often place an enormous strain on both the building fabric and its historic contents. In particular, the impact of heating and ventilation on sensitive surfaces, such as wall paintings, is little under- stood by many charged with the care of such buildings and, as a result, damage can often be caused by actions intended to alleviate it.

In order for church buildings to maintain their proper function, it is essen- tial that the historic fabric remains in good condition. Equally, for the building to remain relevant, visited, and funded, it is important that the conditions within it are conducive to its working nature. It is the role of the conservator to advise on the ways in which conditions can be achieved that are acceptable to the people using the building, as well as being suitable for the conservation of the build- ing fabric and the objects displayed within it.

Introduction

It is an often-quoted maxim of Paolo Mora, the eminent Italian restorer, that the two things that cause damage to wall paintings are man and moisture. While conservators and building professionals can often control moisture through technical means, to control man, and the effect that he has on the historic environment, is a rather more complex task.

It is this aspect of conservation – the effect that people have on the historic buildings in which they live and work – that is to be addressed here. Although this paper focuses on the effects of building use on the structural fabric and internal surfaces, in particular wall paintings and architectural polychromy, of churches and cathedrals, many of the principles involved are relevant to all manner of historic buildings.

The problem

Churches exist for the worship of God and, in many cases today, they have a secondary function as a social hub for the community. Although this may seem obvious, it is a fact that often appears to be forgotten when considering the conservation of the historic fabric or objects within the churches. (How many conservation plans, for instance, mention the word 'God'?) However, understanding this fact is critical to successful long-term conservation.

The requirements of the people using these buildings are often at odds with those of the conservation community whose role it is to maintain and conserve the building fabric. The expectations of modern congregations and visitors, in particular the levels of comfort heating that are now considered desirable, are very different from those of only a generation ago, and can place an enormous strain on both the building fabric and its historic contents.

In the view of the author, the job of the conservator is to bring the historic building to the next generation in a stable and sustainable condition. We often forget that, in the context of the history of the buildings that we treat, conservators are a tiny irregularity. The day-to-day, or rather decade-to-decade, care is provided by the people who own or maintain the building. Therefore, in order to ensure that the building is properly conserved, it is essential that we persuade these people of the validity of our advice. Because, if once our collective backs are turned, they do not feel inclined to implement our advice, then what use is it? As a profession, we are extremely weak in this regard. How many conservation courses, for instance, have a module entitled 'dealing with the client'?

It is certainly true that there has been a considerable development in preventative conservation over the past decade. In this context, preventative conservation means the identification and treatment of causes of deterioration, before severe damage develops, rather than the treatment of the symptoms after the damage has already taken place (an approach that has characterized so much historic conservation and restoration). A key to this change of focus has been the fact that a preventative approach has been recognized as worthwhile by some, although by no means all, of the funding bodies. However, despite these positive developments, the basic problem remains. People wish to use historic buildings in a way that they have never been used in the past, and this often comes into direct conflict with the requirement to conserve the historic fabric and contents in their present state.

Specific issues

In order to explore this dichotomy, and to look at the ways in which these problems can be overcome, some of the specific issues that commonly cause difficulties in historic churches are addressed below.

Modern heating

Perhaps the most widespread problem is that of the internal environment and, in particular, the effects of modern heating systems. In general, people who use churches would like them to be warm and dry during services, but, for reasons of cost, they do not wish to maintain these conditions at other times. As a result, heating is turned on for two or three hours on a Sunday morning and turned off for the rest of the week. Although this approach is now widely understood to be damaging to the building fabric, it is one that is still maintained in thousands of churches across the country. The microclimatic effects that such heating patterns cause, as well as some of the more common misconceptions associated with them, may usefully be reviewed.

The sudden increase in air temperature that takes place when heating is turned on, will usually cause the relative humidity (RH) to drop. Typically, in medieval churches, the change in RH will be in the order of at least 10 per cent or 20 per cent, and will cross the band of deliquescence of numerous salts and salt mixtures that are found in original and added building materials. This causes movement and crystallization of the salts and resultant damage to the building fabric and, in particular, to the most sensitive surfaces such as wall paintings and architectural polychromy. The reduction in air temperature, when the heating is turned off, allows the RH to increase and the salts go back into solution. Each cycle of heating repeats this process, causing damage to the fabric on every occasion.[1] Although this is a huge over simplification, the mechanism it describes is widely known about by historic building professionals.

What is less widely appreciated is the effect of an increase in temperature on the absolute humidity, that is the actual volume of water in the air. The fabric of most medieval churches acts like a sponge and absorbs high levels of moisture from both water vapour in the air and liquid water in the building structure. When the air temperature is raised, some of this moisture desorbs or evaporates, increasing the actual moisture content in the air. This increase in the volume of water in the air means that when the heating is turned off, causing reduction in air temperature, the RH will increase to a higher level than before the heating was turned on. This is usually exacerbated by the fact that the heating has been turned on for a service, when additional water vapour will have been created by the breathing and perspiring of the congregation, as well as evaporation from wet coats and umbrellas.

Although air heats up and cools down relatively swiftly, many building materials have a higher thermal lag and react more slowly. The warm wet air created during the service comes into contact with the cool surfaces, reducing the air temperature sharply. The cool air can no longer support the high level of water vapour causing it to condense on to the cold surface. In churches, much of the resultant condensation is invisible as it is absorbed directly into the porous walls, but the effect on the dissolution and movement of soluble salts is none the less dramatic.

So here, in a nutshell, we have the problem caused by short-term heating. We understand it, and we understand its long-term implications for the building fabric, but does the parish? Usually not. What is more worrying is that many building professionals appear not to understand it either. As a result, new heating systems are regularly installed in churches, which exacerbate rather than reduce this problem.

How then do we overcome these problems? The first thing is to recognize our lack of control over the long-term management of the buildings in question. In general, the situation for churches is very different from, say, National Trust properties, where there are recognized protocols and the need for environmental control (often in the form of conservation heating) is widely understood. If the congregation in a church feel cold, they will simply override any heating controls in order to make themselves feel warm. If the price of oil goes up, or the pensions market goes down, the money will not be available in the parish to pay for, what is often perceived as, 'unnecessary heating' and our conservation heating system will be turned off. Nor should we ignore the financial implications of falling attendance figures in churches that will inevitably reduce parish incomes, placing further stress on funds for conservation heating and other control systems, which are often regarded as luxuries. These are real issues and, as responsible conservators, we cannot afford to ignore them when we are designing conservation strategies. This is not to suggest that active environmental controls such as 'stabilizing' heating have no place in churches. Simply, we should recognize that, in many cases, such systems would not be used in the way that we intended.

In looking at heating strategies that will work, we should first address the clients' perception of what they think they need. The lowest impact on the environmental conditions can often be achieved with carefully thought-out localized heating. However, people generally expect the conditions in their church to be as similar as possible to their living rooms. So, when they enter a church and the air temperature in the aisle is low, they insist that the church is cold, never mind the fact that the temperature in the pews, where a carefully-designed localized heating system is in operation, may be entirely acceptable. In addition, there is the question of how people distribute themselves around a church. When working in churches, it is common to see a congregation of 20 people spread themselves around a church built for 200. The implications for heating (and heating costs) are obvious.

Many of the technical problems associated with different heating systems have been addressed and information is readily available on such matters. *Heating Your Church* by Bill Bordass and Colin Bemrose is a particularly good introduction for the technician and the layman alike.[2] However, in reality, the problems are not simply technical, but ones of perception. It is the job of the conservation professional to overcome these issues just as we would any other technical conservation matter.

So, making people warm in church, without damaging the historic fabric, involves a very wide range of issues that are not the domain of any one professional advisor. In some cases, one does see imaginative and well thought through solutions to the problems, but in many more cases the discussion is largely about flow temperatures and thermostats. If we are to persuade our clients of the importance of conservation issues related to heating, we must ourselves have a unified approach to the issues involved. This, in turn, means that conservators should be involved in the initial design of any project in just the same way as the architect, heating engineer, and planning officer.

Ventilation

Heating is one of the principal artificial factors affecting the internal microclimate in churches. The other is ventilation. Ventilation is both cheap and easy to implement and, therefore, it is recommended by church architects on a regular basis. The reason usually given for increasing ventilation is to 'dry out' a church. But is this desirable and is this what will be achieved?

Let us first consider the question of desirability. A significant reduction in the water content within the building fabric will inevitably lead to the crystallization of hygroscopic salts, resulting in damage to the fabric. If this is a single event, followed by a stable dry environment, then it might be regarded as worthwhile. However, drying caused by ventilation is usually short term and intermittent and, as a result, repeated cycles of the dissolution/crystallization process take place causing continual damage. Also, recent research has shown that even a mild increase in the level of air movement over the surface of a wet porous object (such as a plastered wall) causes a huge increase in the level of evaporation and, by implication, salt crystallization.[3]

More significantly, if there is water in the building fabric, is ventilation and surface drying the way to treat the problem? Why is the water there in the first place? Often it is as a result of a failure in the building envelope or rainwater disposal system. If this is the case, then a reduction in moisture content through ventilation will be of questionable value. Rather, it will be necessary to repair the damage to the fabric; in other words, treating the source of the problem and not the symptom.

So, the question of the desirability of ventilation is less straightforward than might at first appear to be the case. Certainly, for the building fabric, while long-term stable drying may have certain benefits, short-term drying is likely to cause further damage.

The question of whether ventilation achieves an overall reduction in humidity levels is also far from straightforward. Ventilation, in this context, means the exchange of internal and external air. In historic churches, a high level of what could be termed 'natural ventilation' occurs as a result of building porosity (i.e. gaps in the building fabric that allow air exchange to take place). Deliberate ventilation, on the other hand, involves the encouragement of air exchange via doors, windows, or other openings that are constructed for this purpose.

On average, in Northern Europe, it is wetter outside than it is inside. It therefore seems remarkable that the idea has developed that the uncontrolled importation of external air will somehow make things drier. What actually happens is that, when it is drier outside (i.e. absolute humidity is lower), the imported air will usually reduce the level of water vapour in the internal microclimate. However, if the external absolute humidity is higher than inside (and in Northern Europe, it generally is), it will make the air wetter. There are, of course, sophisticated mechanical systems that will assess and compare the internal and external conditions and allow controlled ventilation accordingly. But this is not what happens in parish churches, where windows and doors are left open or closed for weeks or months at a time.

Even when a more careful approach is taken, the results are rarely much better. Human beings are notoriously bad at judging minor changes in RH, being far more sensitive to fluctuations in temperature than in moisture. As a result of this inability to judge moisture levels, and the very common misconception that warm means dry, controlled manual ventilation tends to consist of throwing open the windows on an occasional warm winter day. On such a day, although the external air temperature may have increased, the absolute humidity will often be high, as a result of moisture evaporating from the ground. This warm wet external air then enters the church and comes into contact with the cold walls, causing water vapour to condense.

This is not just a theoretical model. Environmental monitoring of churches, where a high level of deliberate uncontrolled ventilation is employed, has regularly shown incidents of condensation associated with ventilation (Figures 1 and 2). Indeed, in many cases, incidents of what can be termed 'ventilation condensation' outweigh those of condensation resulting from the inappropriate use of the heating system.

This appears to fly in the face of the extensive anecdotal evidence of ventilation causing a building to dry (this is something that the author has often found puzzling). However, it is often the case that ventilation is implemented as part of a wider programme of building fabric repairs, which address damage to the building envelope and the rainwater disposal system. Therefore, it is possible that the perceived drying caused by ventilation has in fact been caused by the control of liquid water entering the building fabric.

However, giving advice that appears to go against people's intuitive beliefs is extremely difficult, particularly if another building advisor is saying that ventilation is so obviously a good thing. Perhaps ventilation should be considered in the same way as heating. Both are powerful tools for environmental control that, if used carefully, can be beneficial. However, if they are used without adequate control, both can cause considerable damage. Therefore, before either ventilation or heating is recommended, the actual effect that they might have should be considered far more carefully than is currently the case. When a new heating system costing £20,000 is proposed, the advisor thinks about it very carefully, weighing up the pros and cons before recommending it. If it were necessary to pay a similar amount for ventilation, and one had to consider the pros and cons to the same degree, would it be recommended as widely? Probably not.

Figure 1 *The fourteenth-century wall paintings at St George's Church, Kelmscott, Oxfordshire which were the subject of an extensive environmental study.*

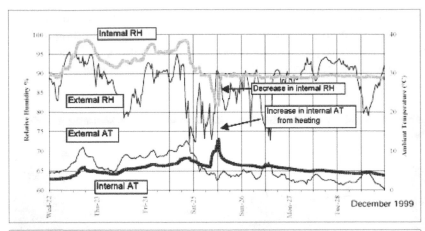

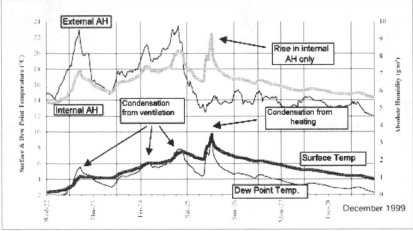

Figure 2 Microclimatic conditions at Kelmscott church in December 1999, showing condensation taking place as a result of external ventilation in the period before Christmas and as a result of heating on Christmas Day.

Alterations to the building fabric

Although these effects can be serious, deterioration resulting from changes in the microclimate is generally fairly slow. Far more immediate is the damage often caused by work on the building fabric, including re-decoration, re-organizing the internal fittings, or the installation of kitchens and lavatories. With certain exceptions, all these changes can be achieved with minimal impact on the historic fabric. However, this relies on careful consideration of potentially sensitive areas of the building at the initial design stage of any project. This is particularly relevant for large projects, such as the addition of lavatories, kitchen facilities, or meeting rooms within the existing church building, where the risk of damage to the original fabric is extremely high. It is therefore essential that the project manager, usually the church architect, recognizes the potential risks at the outset and takes advice from the relevant expert (be that a wall painting conservator or an archaeologist).

With the redecoration of medieval churches, it is now a common recommendation by the Council for the Care of Churches, English Heritage, or the Diocesan Advisory Committee that a preliminary examination is undertaken by a conservator, in order to establish if any wall paintings survive and to recommend strategies to prevent damage during the building work. While this is not yet a universal approach, it is an extremely positive step and should be actively encouraged. As well as protecting wall paintings from inadvertent damage, it can also be very cost effective when one considers the expense that is incurred if wall paintings are discovered in the middle of a building project.

Day-to-day maintenance
Although it tends to be very serious when it occurs, the number of incidents of large-scale damage caused by decoration or structural alterations is limited due to the planning regulations and faculty requirements. Far more common is the non-malicious damage that is caused on a day-to-day basis due to simple ignorance of the historic fabric. Despite the extensive material published by the Council for the Care of Churches, unnecessary damage of this type continues to occur.

A typical example is found on a very fine thirteenth-century painting of St Thomas Becket in a small church near Cambridge. Located in a niche on the east wall of the nave, the painting showed signs of deterioration and paint loss on its lower half, only some of which appeared to be associated with the moisture patterns observed elsewhere on the wall. It was only when the church was visited at Christmas and a vase of holly was found sitting in the niche, pushed hard up against the painting, that the cause of deterioration became clear.

A similar case is that of a very important and unusual sixteenth-century painted stone screen in a large parish church. The screen separates the chapel, dedicated to the patron, from the chancel. While the chapel side of the screen is finely painted with grotesques and polychrome decoration, the chancel side, which was originally very simply painted, had lost much of its polychromy, in particular on the upper tracery. During the initial survey a number of complex theories were examined as to why this pattern of preferential deterioration had developed. It was only some years layer, while carrying out a follow-up survey, that one of the well-meaning church cleaners was found scrubbing down the chancel side of the screen with a big broom 'to get rid of all that dust'.

Far more worrying was the fate of the west end of the screen, which retained much of its very fine painting. Because the screen forms a tight space by the organ, it had become used for storing brooms, vacuum cleaners, bits of old wood, and various other implements, which chipped away at the paintings whenever they were moved. This was pointed out to the church wardens and it was recommended in the report that the practice was stopped. On returning several years later for the follow-up survey, it was found that not only had the advice been ignored, but further material had been jammed in and the damage had worsened. When this was brought to the attention of the church warden, he explained that there was no other convenient place for storage.

The particularly depressing aspect of this case was that considerable time, effort, and money had been spent on behalf of the parish, and recommendations that would have cost no money at all to implement were ignored. As a result, further irreversible damage has occurred to a highly important work of art. Were this a health and safety issue, where professional advice had been wilfully disregarded and injury had occurred to visitors, legal proceedings would follow. In conservation, such things are almost unheard of and as a result this sort of thing regularly happens with no consequence.

Maintaining historic contents will always have risks attached and the simplest actions can cause significant long-term damage. Scrubbing medieval floor tiles, polishing brass plaques on painted walls, cleaning stained glass windows, chipping candle wax off alabaster monuments, even sweeping the floor next to pieces of medieval sculpture all have risks attached to them. The job of conservation advisors is not to stop house-keeping functions from happening, but simply to direct them in a way that will not cause unnecessary harm to the historic material involved.

Building maintenance

Although incorrect housekeeping measures can cause damage, this is rarely as serious as that which can be caused by a lack of, or incorrect, building maintenance. At one end of the scale are the major problems affecting the building fabric. Structural instability, failure of the roof, damage to windows, and large-scale deterioration of pointing are among the most serious issues that tend to be encountered. The system of quinquennial inspections by the church architect or surveyor means, however, that in many cases the problems are identified and acted on before significant damage takes place. Neverthless, in many other cases, a lack of will or funds means that the advice of the architect or surveyor is ignored or acted on far later than has been recommended and further damage occurs.

At the other end of the scale are the minor maintenance issues that take little time or money to implement but are often ignored, sometimes with disastrous consequences. Perhaps the most common problem that results in the most serious damage is the failure to maintain the rainwater disposal system. Blockages of gutters, downpipes, and drains result in more damage to wall paintings than almost any other single cause. The cost of commissioning a local builder to undertake a regular programme of gutter maintenance might be a couple of hundred pounds per year. The cost of the conservation of seriously damaged wall paintings or wet rot in the roof timbers might well run into tens of thousands of pounds.

Conclusions

How then does the conservation profession ensure that buildings are successfully conserved, without preventing them being used for their correct working function?

In fact, most of the requirements of a working church can be achieved without damage to the historic fabric. What are the reasons for serious damage continuing to take place on such a regular basis? Generally, such damage falls into two categories – wilful damage and inadvertent damage. Wilful damage, in this context, means damage that takes place as a result of a deliberate refusal to seek, listen to, or act on expert advice. Inadvertent damage, on the other hand, tends to occur as a result of incorrect expert advice or simple ignorance. Fortunately, most of the damage encountered falls into the second category, although there are still a significant minority of cases where this is not the case.

If we are to attempt to reduce the level of inadvertent damage, the first stage must be to make sure that the expert advice provided to parishes is as accurate as possible and that they do not receive conflicting advice from different advisors. This demands a level of co-ordination between professional advisors that is, at present, uncommon. If, for instance, a church architect is asked to advise on installing or changing a heating system in a church where wall paintings are present, then it is essential that advice is sought from a wall paintings conservator on the impact that this will have on the paintings. Similarly, if a wall-painting problem relates to the building structure, then the conservator should liaise closely with the architect or surveyor. Only in this way will suitable and co-ordinated advice be available to the client, who has no choice but to rely on the accuracy of the information provided by the professional advisors.

The second part of the solution is to ensure that the custodians of the building, in the case of churches this generally means the parochial church council (PCC), have the information about how to care for their building without causing damage. Members of PCCs are usually volunteers and, understandably, do not take kindly to being instructed by 'experts' on how to sweep floors or carry out other basic housekeeping tasks. However, most members of PCCs care very much about the church buildings for which they are responsible and are happy to take advice if it is offered with some tact.

How, then, is such information to be provided? This, in the opinion of the author, is the key to solving much of the problem. Although there is a great deal of extremely good advice available, it often fails to reach those who are actually undertaking the day-to-day care and management of the building. How many volunteer church cleaners, for example, have had any instruction on how to carry out their tasks without causing damage to the historic objects? What is needed is a system by which advice on good practice can be provided to a person within the parish, who can then disseminate it more widely. After all, volunteer archaeologists expect basic instruction before being allowed to participate in a project. Should not similar information be offered to church volunteers?

There remains the difficulty of dealing with damage caused by wilful disregard of advice. In theory, there is a legal framework that provides sanctions when serious damage has been caused, but in practice prosecutions are rare. In cases where minor damage is caused, there appears to be little or nothing that can be done and it is left to the conservation advisors to persuade the parish to implement their advice. Regrettably, given that follow-up surveys are rare, this often has little or no effect.

Although it is often difficult to bring together the working needs of the building and its conservation requirements, we should bear in mind the consequences of failure. If churches ceased to be working buildings, they would lose their purpose and become merely museums. As a consequence, they would lose the people who, at present, freely give their time as custodians and who provide much of the funding for their care. No amount of state aid would make up for this and, in conservation terms, the resulting lack of day-to-day care and maintenance would be disastrous. Therefore, while it is our responsibility to ensure that the historic fabric is properly conserved, it is also our responsibility – wherever possible – to do so in a way that will allow those who use the building to achieve their requirements. Of course, there will always be some cases where the changes that a parish wishes to make will be in direct conflict with the protection of the historic fabric, and in such cases the changes should be blocked. Nevertheless, in the majority of cases, with correct and co-ordinated advice from their professional advisors and careful implementation of the work, most of the building use requirements can be met with an acceptably low level of impact on the fabric.

Biography
Tobit Curteis BA (Hons), Dip Conservation (Courtauld Institute)
Tobit Curteis holds a BA in history of art (University of Warwick, 1988), a postgraduate diploma in the conservation of wall paintings, Courtauld Institute of Art/Getty Conservation Institute (1991), a post-diploma internship in Rome and Florence for the Courtauld Institute of Art (1992), and has been a private conservator in the United Kingdom since 1992.

Notes
1 In recent years there has been extensive research into the behaviour of salt mixtures in historic building materials, and this has shown that the situation is far more complex than has hitherto been understood. However, the basic principles of fluctuations in heating, causing salts to go in and out of solution, remain the same.
2 Bordass, W. and Bemrose, C., *Heating Your Church*, Church House Publishing, London (1996).
3 Pender, R., 'Towards Monitoring Moisture Movement in Support Materials of Wall Paintings', *Proceedings of the 6ᵗʰ International Conference on Non-Destructive Testing and Microanalysis for Diagnostics and Conservation of the Cultural and Environmental Heritage*, Rome, 1999, pp. 831–41.

Further reading
Council for the Care of Churches, *How to Look After Your Church*, 3ʳᵈ Ed., Council for the Care of Churches, London (1991).
Council for the Care of Churches, *A Guide to Church Inspection and Repair*, 2ʳᵈ Ed., Council for the Care of Churches, London (1995).
Curteis, T., 'Wall painting conservation in England: A perspective from private practice' *Conserving the Painted Past, Developing Approaches to Wall Painting Conservation*, postprints of the English Heritage conference, December 1999, James & James, London (2003).
Burman, P. (Ed.), *Treasures on Earth, A Good Housekeeping Guide to Churches and their Contents*, Donhead, London, (1994).
Staniforth, S. and Sandwith, H., *The National Trust Manual of Housekeeping*, The National Trust, London (2002).
Bordass, W. and Bemrose, C., *Heating Your Church*, Church House Publishing, London (1996).
Massari, G. and Massari, I., *Damp Buildings Old and New*, ICROMM, Rome (1993).
Thompson, G., *The Museum Environment*, Butterworth-Heinmann, London (1994).

When Conservator Meets Architect and Engineer

SARAH STANIFORTH AND KATY LITHGOW

Abstract

Environmental monitoring in historic buildings has shown that the summer climate of the United Kingdom is usually benign for the collections housed within them. Solar gain is sufficient to reduce the naturally high relative humidity of our maritime climate below the level at which damp-related problems are triggered. Internally, however, comfort-heating levels for human occupants can produce relative humidity levels that are disastrously low or dangerously fluctuating, not only for collections but also for fixtures and the building structure. Conversely, unheated buildings suffer from mould, rot, insect attack, and metal corrosion. The National Trust creates constant humidity at levels below the threshold for mould growth (65 per cent) by installing conservation heating systems controlled by humidistats.

To ensure a holistic approach to preventive conservation, other services are often upgraded when conservation-heating systems are installed, along with maintenance of the building fabric. The associated building work threatens collections and historic interiors through physical damage, dust, and theft. These risks are controlled through planning by a multidisciplinary team underpinned by good project management. Planning minimizes physical disruption, and programming ensures sufficient time for storage and protection as well as installing and commissioning new systems. Risks are controlled by specifying working methods, providing storage, designing protection, and employing specialist staff to ensure these measures are implemented.

Introduction

This paper will consider three areas:

- The balancing act that has to be struck when conserving collections of fine and decorative arts that are still housed in the historic buildings for which they were made;
- The importance of risk assessment and risk management in caring for collections and the buildings in which they are housed; and

- The multidisciplinary teamwork that is needed when planning and managing projects to renew or replace services in historic buildings.

Training for conservators and other professionals working in historic buildings

The majority of training courses for conservators are designed to equip students with knowledge about the conservation of individual objects. At the end of their training the conservators emerge well prepared for taking up jobs in museums. But every year a handful of newly-qualified conservators begin work with organizations such as English Heritage and The National Trust, where collections are housed in historic buildings. The National Trust employs over 30 conservators and a much larger number of conservation advisers and freelance conservators. For us, the conservation of an object must be carried out with understanding of the context in which it is displayed. Before decisions can be made about any conservation treatment, the historic, social, aesthetic, and cultural significance of the object and its surroundings must be considered.

An example of the way in which the treatment of a painting is affected by its surroundings can be illustrated by the Long Gallery at Hardwick (Derbyshire), where a fine collection of family portraits hang in front of a set of tapestries telling the story of Gideon (Figure 1). Four hundred years of light exposure have resulted in the vibrantly coloured natural dyes fading to dull shades of brown and blue. By contrast, the colours in the paintings have survived quite well. A conservation treatment of the paintings that sought to remove discoloured varnish to recover the original pigment colours might result in an undesirable imbalance between the paintings and the tapestries.

The different thought processes required of a conservator when considering the remedial treatment of an object in a historic house are even more evident when preventive conservation measures are planned for the preservation of collections. In a historic house collection, the building itself is the largest and often the most important artefact that must be preserved. Environmental control systems that are appropriate in purpose-built museums, such as air conditioning systems that are designed to deliver clean air at a specified temperature and humidity, cannot be installed in historic buildings without an unacceptable level of damage to the structure being incurred when holes are made for ductwork to pass through.

Since the majority of conservation training courses are not designed to teach the special measures that need to be taken for collections in

Figure 1 *Long Gallery, Hardwick Hall, Derbyshire. (National Trust Photographic Library/Andreas von Einsiedel)*

historic buildings, The National Trust has had to develop its own training for conservators. This is provided by specially-designed internal courses, and through training and development posts for recently qualified conservators.

In the same way that conservators have had to develop their skills and knowledge to adapt to working in historic buildings, we notice that there is a similar need for architects and engineers to develop their sensitivity to historic buildings. There are several courses that architects can take in historic building conservation, but much less provision for engineers. Many engineers have developed their knowledge about working in historic buildings by working alongside other professionals in complex building projects. There have been suggestions for courses in which professionals concerned with the conservation of historic buildings can study alongside each other, and benefit from a broad range of experience. Any course that manages to attract a multi-disciplinary body of students to study historic building conservation is particularly welcomed by The National Trust.

Environmental management

As an illustration of The National Trust's approach to preventive conservation in historic buildings, we could look at priorities for reducing risks to collections.[1,2] There are the obvious risks caused by catastrophic events: fire, flood, and theft. The floods in central Europe during the summer of 2002 are a warning for us to be aware of the increased risks of flooding resulting from climate change – not only of the risks posed by rivers, but also of rising water tables and heavier rainstorms. We are beginning to see signs of rainwater goods not being able to cope with extreme rainfall. There are also cumulative risks posed by physical wear and tear, light, and the effects of the wrong temperature and relative humidity.

Environmental monitoring in historic buildings has shown that the summer climate of the United Kingdom is usually benign for the collections housed within them. Solar gain provides sufficient gentle warming to reduce the naturally high relative humidity of our maritime climate below the level at which damp-related problems, such as mould growth, are triggered. The substantial construction of many historic buildings means that the insides are well insulated and buffered from temperature and relative humidity fluctuations, although opening doors and windows can create problems with condensation at any time of the year.

The internal climate during the winter is a different story. Buildings that are heated to levels comfortable for their human occupants can experience relative humidity levels that are disastrously low for collections and also for fixtures, such as panelling, and for the structure of the building itself, including timber frames. Large swings in temperature and relative humidity are caused by turning heating on and off on a daily cycle. Conversely, unheated buildings experience a multitude of damp-related problems including mould, rot, insect attack, and metal corrosion.

Unheated buildings can also experience condensation events, which are often misdiagnosed as water infiltration.[3,4] Conservation heating (i.e. heating systems controlled by humidistat to create constant relative humidity at levels below the threshold for mould growth at 65 per cent) has been installed in many houses.[5] Achieving recommended relative humidity levels through temperature alone occurs only at the expense of human comfort. This is, however, a realistic option for organizations such as The National Trust whose houses are closed to the public from the end of October to the beginning of April.

Conservation heating systems are often installed as part of a much larger project to upgrade all services, including security and fire detection systems, as well as maintaining the fabric of the building, to ensure a holistic approach to preventive conservation. Where possible, systems are selected that minimize interference with the fabric of the buildings. Wireless radio telemetric systems have been successfully used for fire detection, and for monitoring and controlling conservation heating systems.[6] The introduction of other preventive conservation measures have less easy solutions, and much ingenuity is required at the planning stage to select and specify solutions that minimize impact on the fabric of the building.

Risk management

The main threats of such installations to collections and historic interiors are physical damage and wear and tear, dust and dirt, and theft.[7,8] These risks occur not only at the site of work, but over the routes providing access into and throughout the building. People are affected by these projects as well as objects. There is a great deal of stress involved for house staff who have to supervise the work as well as carry out their normal jobs, and in some cases live on the work site. As contractors work from Monday to Friday with hours that range between 7.30am to 5pm, house staff have to be on site not only during the same hours to supervise their work, but also an hour before to open up, and an hour afterwards to lock up after them.

Planning
In our experience, these risks are controlled through planning and programming by a multidisciplinary team involving engineers, architect, conservators, curators, building staff, property staff, and contractors. We find that this coordination resolves any potential conflicts in advice. The key to a successful outcome is in the project management, with good communication through regular team meetings. Planning should start two years before the contractors come on site because of the need to budget in advance, the necessity of applying for listed building consent(s), and advance warning needed to book freelance conservators and specialist removals staff.

Programming
Designing a practicable programme for the contract involves not only the time needed for the contractors to do the work, but also planning time, storage of moveable objects, protection of vulnerable surfaces, moving stores between phases of work, redecoration, cleaning the house, commissioning new systems, contingency periods for the contractor, and finally

returning and, where necessary, re-displaying the contents. Without planning, the timetable focuses on the contractor's work, and may leave insufficient time for safe packing and reinstatement of the collections, so enabling the work to happen safely and speedily, and checking that it has happened successfully.

Sometimes people only apply this thinking to major projects, where houses are closed for more than the usual five-month winter period, and the contract value is in excess of £100,000. But minor projects can be equally disruptive, and if a keen planning eye is not kept on them, they can grow in number as advantage is taken of the existing disruption to add more 'minor' projects, resulting in too much being packed into too short a time with too few resources. However large or small the project, appropriate planning is needed to ensure that there is the capacity to do the work and protect the building and its contents properly.

Design systems to reduce impact
Effective planning can minimize damage by identifying the least disruptive routes and working methods before the contract starts; for instance, taking up floorboards that have been taken up before. As discussed above, conservation heating systems now minimize damage to the fabric of the building through the use of free-standing electrical heaters, telemetric monitoring, and adaptation of original systems, but some impact is inevitable when wiring and pipework is installed.

Flameproof mineral insulated cable, commonly referred to as MIC or Pyro, is one of the most high impact elements of building work. This is now used universally to rewire our houses because it is fireproof and inedible to rodents. Six miles of this cable were installed in Nostell Priory (West Yorkshire). However, it is rigid and therefore more difficult to handle than pvc cable, requiring special equipment to cut and straighten, more room to fit into wall and floor voids (especially to get around corners), and is therefore more disruptive to building fabric.

Changing software standards mean that control boards are out of date more frequently. Year 2000 compliance meant we had to replace installations that were only seven years' old, and there is no reason to assume that such in-built obsolescence will not recur. Sometimes it seems that the tail (protective building services) is wagging the dog (the building).

There are some developments that may encourage a move towards a longer lifespan for building services. Local interfaces known as RIOs (Remote Input Output modules for the Ademco Microtech Galaxy range of intruder alarm panels) should enable sensors to be added or upgraded without taking their wires all the way back through the house to the central control board, reducing the disruption to the room or circuit of rooms that the RIO serves. Nostell's 9.6 km of MIC cable was less than the 50 km laid at Ham House (London), perhaps showing the benefit of using RIOs. The equipment is, however, housed in boxes some 150–200 mm square, and may need to be concealed by being cut into the building fabric, behind shutters and panelling, or beneath floors.

In the same way that methods of achieving environmental control have moved away from complex air-handling systems to simpler hard- and surface-wired, remotely-monitored heating systems that re-use original installations, similar requirements should be made of other building services. These could include simplifying equipment and materials, and ensuring longevity, not only through the materials used, but also through easy maintenance and upgrading. This would make building services far more sustainable.

Storage

The National Trust protects the contents and fabric from theft, dust, and mechanical damage in several ways. Portable objects are moved out of the work area into a safe store. This also provides the contractor with safe access to the room, which speeds up work.

The scope of the building work dictates the amount of storage and protection required and thus how long the house has to remain closed to the public. There are several options based on storage within the house, on-site within outbuildings, or off-site in commercial storage, used separately or in combination, with a range of costs. These costs increase with the distance that objects have to travel to store, and the number of objects and length of time for which they are stored.

An analysis of storage costs in the 1990s identified a range of costs. Storing contents within the house costs between 3.6 per cent and 7 per cent of the contract sum, typically between £6,300 and £16,500. Storing them in outbuildings on the estate, upgraded to provide improved security and environmental control, costs between 24 per cent and 30 per cent of the contract sum, or £50,000 to £500,000. Using commercial stores costs between £100,000 and £500,000, of which hiring the store and its equipment costs between £25,000 and £30,000. Hiring our own stores becomes more cost effective as the time of storage increases, spreading the costs of equipping the stores over time.

If the costs of *in situ* protection and temporary services (such as security, fire detection, and environmental control) are added to storage costs, the total budgets required to protect contents and fixtures range between 5.5 per cent and 17 per cent of the total contract sum.

Deciding on the appropriate storage solution involves balancing the costs of storage and protection against contract costs and visitor income. A decision to store contents on site may result from deciding that the lower costs of such storage are better value despite the increased time the contractor will need to work around stores, and the loss of visitor income if prolonged closure is necessary. The decision to store contents off site may result from comparing the higher costs of storage with the benefits of achieving building work in a much shorter time frame (usually within the five-month closed season) and increasing visitor income by reopening earlier.

Working practices

We try to remove the risk of damage at source. Safe working practices for the contractors are identified and specified in contract documents, such as: using vacuum cleaners and dust tents at the source of dust creation; specifying ways of carrying and storing equipment and materials to avoid impact damage; and methods of removing floorboards and panelling. Contractors are given induction training by National Trust staff, not only to identify fire exits, but also to explain the house and the nature of its sensitive areas. Explanatory notices on protection explain what is underneath surfaces, and also on floors over fragile plaster ceilings or walls behind fragile decorative surfaces. These measures are identified through internal guidelines,[9,10] and national guidance documents.[11,12]

In situ protection

On the basis of these working practices, we design in situ protection for the most vulnerable areas, which are beside work sites and access routes. Where the room affected is large, it can be cheaper and less risky to box in the contractor than protect an entire room, providing health and safety legislation is complied with. Here, the intention is to remove the dust at source by the use of vacuum cleaners, sometimes in combination with dust tents. In fact, dust extraction within the tents becomes an important part of the contractors' protection as well as that of the house.

A range of low-cost materials that conform to conservation criteria in terms of chemical stability and environmental performance are available (such as Tyvek, downproof cotton cambric, and acid-free tissue), which provide the range of varying weight and flexibility necessary for each different protection situation. We avoid damaging historic fabric by fitting protection with ties, wedges, bracing, and pressure fixing, so minimizing any mechanical fixings and avoiding adhesives.

At Nostell Priory, the most vulnerable areas included a number of pier glasses that were beside the main wiring route running in front of the windows. In the State Bedroom, the Chinoiserie pier glasses are flanked by green lacquer furniture and hung against hand-painted Chinese wallpaper, all provided by Chippendale in the 1770s. The glasses were protected against mechanical damage by rigid screens, which had to provide full access to the floor void (Figure 2). The screens were therefore canted off the wall, by being fixed at the top into the unfinished upper sides of the wooden pelmet boards flanking the glasses, allowing them to be held in against the bottom of the wall by a softwood 'clamp' around the pier. This clamp was fixed by being wedged into the window reveal on either side of the pier, protecting the adjacent painted wood and plaster with polyethylene Jiffy Foam padding. Hard protection was provided at the bottom, where the contractors were working, by sheets of Medite, a low formaldehyde-content medium-density fibreboard. Lighter-weight

Figure 2 In situ protection at Nostell Priory. (National Trust Photographic Library/Katy Lithgow)

protection was provided above, where the risk of mechanical damage was less, by Antinox, a fluted polypropylene flame-resistant board.

Conservators design these protection systems in conjunction with the carpenters who fit them. Thus, they are dedicated to identifying practical and ingenious solutions to provide contractors with safe access.

Staff

Project conservators are employed to assist with planning and executing storage and protection, ensure that it is well maintained and adapted as required, supervise contractors, oversee commissioning, and organize the cleaning of interiors and return of the contents. Sometimes accompanied access and supervision is less risky and cheaper than installing vast amounts of protection or moving large collections of fragile items where work is limited. Project conservators are especially valuable where house staff numbers are limited. It is vital for there to be one person in the role, to maintain continuity, and to avoid delays, poor communication, and the possibility of different people being played off each other. It is hoped that these project posts will provide career opportunities for training positions for conservators that we are currently recruiting, as well as continuing to provide experienced freelance conservators with an opportunity to develop their experience of preventive conservation and knowledge of The National Trust.

Specialist staff, such as freelance conservators, are needed to move particularly fragile objects, while larger-scale moves that are beyond the capacity of the house staff, may be undertaken by removal companies working to Trust standards.

If they are willing, part-time house staff may have their hours re-arranged or may be employed for longer to provide full-time cover during the working week, especially during the packing, protection, cleaning, and reinstatement periods.

Where there are no direct-labour sources, or because of tendering procedures, the tried and trusted builders who are familiar with the property may not be awarded the contract. Project joiners trained in conservation methods are sometimes used to open up the building and minimize damage.

Public access

In the interests of maintaining public access, particularly where public money is involved in new acquisitions, we are now trying to evolve ways of enabling public access to the houses whilst re-servicing and building repairs are carried out. Whilst the conservation mind can feel perfectly happy that thousands of pounds have been spent and the house closed for a long time with no resulting visible evidence of the work carried out, we now have to explain how we got there. At Nostell Priory we arranged behind-the-scenes tours, media coverage, and an exhibition explaining the work that had taken place and thus the reasons why the house had been closed for eighteen months.

In addition, there is concern that prolonged closure may weaken our visitor base in the face of increasing competition from other local attractions. For this reason, projects now tend to include refurbishment or opening of new rooms on the visitor route as a temptation to new visitors. However, we should not underestimate the sophistication of many visitors, whose own comments reveal their ability to perceive and appreciate sensitive repair.

The planning surrounding one of the Trust's most recent acquisitions, Tyntesfield (Avon), is underlain not only by the need for conservation standards in repair and display, but also the maintenance of public access throughout the project (Figure 3). The building fabric seems to be in a remarkably good state of repair compared to other houses like Calke Abbey (Derbyshire) and Chastleton (Oxfordshire), but does suffer from a high level of insect infestation due to high levels of moisture in the wood. It has a sophisticated heating and ventilation system that will be re-used on the ground floors to provide conservation heating. It will be our first opportunity to really allow the interaction between conservator, architect, and engineer to be seen by the public.

Conclusion

It has taken a long time to gain recognition that the welfare of the collection and fabric must not be compromised by work designed to safeguard them for the future, and that investment in their protection is a legitimate cost and planning concern.. Conservators have to adopt planning, management, and business skills in order to convince project teams of the legitimacy of their demands. The arguments have to be rehearsed for each project as staff, consultants, contractors, and the scope of the work are often different for each one.

The key to reconciling the apparent inevitability of damage to a building by the very work that is intended to safeguard it, is a holistic approach that recognizes that intervention needs to be kept to what is necessary to safeguard the building from man-made and natural disasters and deterioration, and to the minimum required to achieve these aims. Damage

Figure 3 *Tyntesfield (Avon). (National Trust Photographic Library/ Andrew Butler)*

can be prevented by planning a project so as to be sensitive to the impact on the building, minimizing the disruption caused by letting building services into building fabric, and recognizing that the investment in the protection of the contents is preventing damage. Thus a conservation approach throughout the whole team can become a tool to manage and control risk.

Ultimately building services themselves become of historic significance. Conservation-heating systems can adapt original and historically significant heating systems. Whilst this undoubtedly reduces the impact on the building fabric by re-using wiring and pipe runs, and indeed finds a use for equipment that may otherwise have to be stripped out, inevitably it raises questions about how working objects can be conserved. This is not the place to discuss how far we should be conserving processes as much as parts. Suffice to say that using a system productively can provide a reason and an opportunity for maintaining it that disuse cannot. If done sensitively and with minimum physical impact, measures devised to provide long-term maintenance can be seen as adding to the meaning and significance of a building, and part of its evolution, rather than as a modern imposition. It is perhaps worth considering how future generations of conservators, architects, and engineers could be adapting our own building services to service their needs, just as we are adapting, and therefore in effect preserving, the building services of previous generations.

Appendix: Suppliers

Antinox   Swiftec, Manchester.
Jiffy Foam   National Packaging Plc, Bristol (assessed as suitable for permanent use by the British Museum, in Lee, L.R. and Thickett, D., *Selection of Materials for the Storage or Display of Museum Objects*, British Museum Occasional Paper 111, British Museum, London (1996), p. 51.)
Medite   Local builders' merchants, names from Williamette Europe Limited, Southend-on-Sea.

Biography

Sarah Staniforth BA, Diploma in Paintings Conservation, FIIC, PACR
Sarah Staniforth read chemistry at St Hilda's College, Oxford, and then studied for the Diploma in Easel Paintings Conservation at the Courtauld Institute of Art, London University. From 1980 to 1985 she worked in the Scientific Department of the National Gallery and since 1985 has worked for The National Trust, advising on paintings conservation and environmental control in houses. In June 2002 she was appointed Head Conservator. She is Vice President of IIC (International Institute for the Conservation of Historic and Artistic Works) and is Chair of the UKIC Accreditation Committee.

Katy Lithgow BA (Hons), MA (Cantab), Dip Cons, CAPT, PACR
After graduating from Cambridge University in History of Art, Katy Lithgow trained in wall-painting conservation at the Courtauld Institute of Art, and undertook an internship at the Victoria & Albert Museum, before returning to the Courtauld to teach. She joined The National Trust in 1991 where she has worked as a conservator in preventive conservation with a particular emphasis on storage and protection, participating in over sixteen projects, and as Adviser for Wall Painting Conservation. In July 2002 she was appointed Conservation Advisers Manager.

Notes

1 Ashley-Smith, J., *Risk Assessment for Objects Conservation*, Butterworth-Heinemann, Oxford (1999).

2 Michalski, S., 'A Systematic Approach to Preservation: Description and Integration with Other Museum Activities' in: Roy, A. and Smith, P. (Eds.), *Preventive Conservation: Practice, Theory and Research: Preprints of the Contributions to the IIC Ottawa Congress, 12–16 September 1994*, IIC, London (1994), pp. 8–11.

3 Massari, G. and Massari, I., *Damp Buildings, Old and New*, ICCROM, Rome (1993) [translated by C. Rockwell].

4 Staniforth, S. and Griffin, I., 'Damp Problems at Cliveden' in: *The Conservation of Heritage Interiors: Preprints of a Conference Symposium, Ottawa, Canada, 17–20 May 2000*, Canadian Conservation Institute, Ottawa (2000), pp. 177–84.

5 Staniforth, S., Hayes, R. and Bullock, L., 'Appropriate Technologies for Relative Humidity Control for Museums Collections Housed in Historic Buildings' in: Roy, A. and Smith, P. (Eds.), *Preventive Conservation: Practice, Theory and Research: Preprints of the Contributions to the IIC Ottawa Congress, 12–16 September 1994*, IIC, London (1994), pp. 123–28.

6 Bullock, L., Hayes, R., Singleton, A. and Staniforth, S., 'Conservation Heating by Cable and Radio', *Museum Practice*, 2, No. 1, March 1997, pp. 71–74.

7 Eremin, K. and Tate, J., 'The Museum of Scotland: Environment During Object Installations' in: Brown, J. (Ed.), *ICOM-CC 12th Triennial Meeting, Lyon, 29 August to 3 September 1999, preprints (volume 1)*, James and James, London (1999), pp. 46–51.

8 Lithgow, K. and Adams, S., 'Monitoring Methods of Dust Control during Building Works', *Views*, No. 29 (1998), pp. 26–28.

9 Lithgow, K. and Lloyd, H., *Storage and Protection for Building Works*, unpublished internal guidance paper (2001).

10 The National Trust, *The Building Manual*, unpublished internal guidance manual (2001).

11 Museum and Galleries Commission (Cassar, M., Osborne, P. and Longhurst, A.), *Working with Contractors. Guidelines on Environmental and Security Protection During Construction Work in Museums*, Museum and Galleries Commission, London (1998).

12 Cassar, M. (Ed.), *Museum Collections in Industrial Buildings: A Selection and Adaptation Guide*, Museum and Galleries Commission, London (1996).

Conserving Cardiff Castle – Planning for Success

JOHN EDWARDS

Abstract

Finding out about problems and their causes is obviously the starting point in progressing with the repair of any building or structure. The House at Cardiff Castle – of fifteenth-century origin, but including Roman and Norman elements and re-modelling by William Burges in the nineteenth century – will be used to describe why it is important to consider not just the building itself, but also the internal decorations, use, furniture, and artefacts. The Cardiff Castle conservation plan and the integration of both technical and philosophical issues will be highlighted. The need for a multi-disciplined approach via a number of distinct processes will be discussed. The challenge of getting all consultants to work together and the means of ensuring that their work culminates in satisfactory conclusions will also be described.

Introduction

Planning for success ultimately means considering all the issues, in all respects, all of the time. This means not only considering the interface between the building and its contents, but also weighing up, applying, and sharing knowledge of different philosophies, technicalities, and practicalities. Here, Cardiff Castle, a scheduled ancient monument and a Grade I listed building, is the vehicle used to show how this can be done.

Historical development

Whilst Cardiff Castle dates back 2,000 years, it is the nineteenth century that shaped not only the Castle of today, but also Cardiff itself. Cardiff was a place that saw many changes in the nineteenth century; it grew from a population of a mere 2,000 at the beginning of the century to a massive 164,000 at the end. The foundation of all this is Cardiff Castle.[1]

The 2nd Marquess of Bute, the early-nineteenth-century owner of Cardiff Castle, was the man behind Cardiff's growth and prosperity. His

Figure 1 *West elevation of the House. (Cardiff County Council)*

wealth was such that when the 3rd Marquess of Bute, who inherited his father's wealth, came of age in 1869, he was reported to be the richest man in Britain, and possibly the world. His wealth enabled him, in collaboration with his architect William Burges (1827–81), to fulfil his high Victorian dream and turn Cardiff Castle into a neo-gothic, medieval fantasy. This included the creation of 15 of the most colourful and ornately-decorated interiors one could imagine, certainly within a castle in Wales – most castles in Wales lie in ruins!

William Burges re-developed and added to a house that was substantially fifteenth century in origin, but with Roman and Norman elements. It had also undergone many alterations in the sixteenth century, and in the eighteenth century had been extended by Henry Holland (1745–1806), when the grounds were landscaped by Lancelot 'Capability' Brown (1716–83).[2]

Burges interiors – considering all the issues

One of the themed Burges interiors is the Winter Smoking Room (Frontispiece), situated within the Clock Tower and completed in 1873. During a survey in 1992 it was revealed that the original decoration had, at least in part, been overpainted (or re-decorated). In such a situation,

Figure 2 Summer
Smoking Room.
Members of the client
team discussing the
project. (Heritage
Lottery Fund)

a number of questions must be asked. Is the overpainting damaging the decoration underneath or could it be protecting it? If we want to remove the overpainting, can this be done safely without removing, harming, or destroying the original paintwork underneath? If we can remove the overpainting safely today, can we also do so in the future, or will it become increasingly difficult to remove as time goes by? If we can remove the overpainting, is it appropriate to do so? For example, will we reveal a complete scheme of decoration underneath or is the only complete scheme that we know about the one that we see presented to us today?

In order to structure decision-making in the treatment of the site, a management process has been developed. This is composed of four main criteria – aesthetics, interpretation, education, and technical necessity and feasibility.

Whilst those who visit the Winter Smoking Room are generally very pleased with what they see, others (such as academics) may be less pleased as the decorative scheme is not entirely original.

Is it right and proper that we should see what could be called an interpretation of a William Burges decorative scheme when there could be a complete original scheme underneath? From an educational perspective, should we not reveal the original scheme in order that the visitor might see what the original scheme would have looked like? Or should we not stress to visitors that by looking at the Summer Smoking Room (Figure 2) upstairs one can see how the decoration in the Winter Smoking Room would have looked had it not been overpainted?

Again, from an educational viewpoint, the original decoration may not be presented but the theme is still present, and has only been simplified. The decoration therefore still retains its intellectual value; its iconography can still be read, understood, and enjoyed.

In addressing the three criteria of aesthetics, interpretation, and education, one can witness a great deal of subjective opinion and not much justification for removing the overpainting. A great deal more objectivity is used with regard to technical considerations. Technical feasibility is not in itself, however, justification for removal of the overpainting. The overpainting is not harming the original underlying decoration, and if anything, is providing a degree of protection. If we do want to remove the overpainting, however, it can be done safely today or in the future.[3]

Application of philosophies
If we want to further explore the feasibility of removing overpainting, then we have to take account of relevant conservation philosophies. The *Venice Charter*, for example, forms the basis of much respected and official published guidance to appropriate works on historic sites around the world. This may help guide us in making justifiable decisions.

Article 11 of the *Charter* indicates, for instance, that there must be 'exceptional circumstances' in order to justify removal of the overpainting and thus revealing the underlying state.[4] The criteria considered above quite clearly fails to provide the 'exceptional circumstances'. The only avenue left to explore is that of significance.

We have to determine whether the significance of the original decoration is greater than the significance of the overpainting. If the latter is poor and seriously detracts from the significance of the original it justifies removal of the overpainting. The most ethical way to treat a historic building is to conserve it and although there are numerous definitions used to describe the meaning of conservation, there is one in particular that comprehensively addresses its significance.

The Australian *Burra Charter* states that:

> Conservation means all the processes of looking after a place so as to retain its cultural significance. It includes maintenance and may according to circumstance include preservation, restoration, reconstruction and adaptation and will be commonly a combination of more than one of these.[5]

This definition of conservation does not, however, mean anything unless the term 'cultural significance' is known and understood, as this is what it aims to retain.

Cultural significance is defined as 'aesthetic, historic, scientific or social value for past, present or future generations'. This indicates (without looking at the meaning of each and every word) that conservation is all embracing, with the ultimate intention of securing the cultural significance of the site and its various elements.

Referring to conservation in such a comprehensive way is both ethical and useful, but it may also be appropriate to bear in mind the meaning of conservation as defined for the care of objects. This will enable matters to be considered that may otherwise be ignored. Simply put, it should be understood that conservation equates to preservation and is the practice of minimum intervention. Restoration involves work that is more extensive than minimum intervention and should only be carried out for compelling reasons. How does one determine what is or is not a compelling reason? This again brings us back to determining significance and importance.[6]

More specifically, where the treatment of painted and decorated surfaces are concerned, a conference of the United Kingdom Institute for Conservation (UKIC) defined conservation as:

> ...the means by which the true nature of an object is preserved. The true nature of an object includes evidence of its origins, its original construction, the materials of which it is composed, and information as to the technology used in its manufacture.[7]

In deciding whether there is justification for the removal of overpainting, we must determine a number of issues. We may understand the true nature but have we got 'exceptional circumstances' and 'compelling reasons' as sufficient justification for removal and will this mean preserving the 'true nature' of the interior? The answer may lie in the relative 'significance value' of the original and its overpainting.

In the Winter Smoking Room, something detracts from the significance of the original decoration – it is not complete. The only complete decorative scheme is the one presented to us today. There is thus no justification for removing the overpainting. It is not a question of minimum intervention, but of no intervention at all.[8]

Conservation and management plan

The *Cardiff Castle Conservation and Management Plan* was finalized in 2000 and, as one would expect, considered all issues. These ranged from issues of archaeology, ecology, wildlife, and conservation to business management and visitor management, together with care and management of collections. The document was produced by a range of appropriate experts in consultation with numerous organizations, including Cadw (Welsh Historic Monuments). All this gave the plan the required degree of credibility.

The *Cardiff Castle Conservation and Management Plan* determined significance values together with degrees of vulnerability – the conservation plan provided much valued statements of significance and conservation policies, whilst the management plan determined the most appropriate means of implementing the policies.

There was determination at the outset that the plan was to be a useful document and a tool for informed and appropriate treatment of Cardiff Castle. It addressed issues, for example, that would ultimately have to be considered when conservation work was being proposed. The plan lived up to these expectations – it determined that the most important elements of Cardiff Castle are the Burges interiors and concluded that there may be occasions where the only way to preserve a Burges interior would be to sacrifice part of an external element. An example of this would be the renewal of a roof covering, rather than repair, when its repair might pose a risk of leakage into the interior. The plan also endorsed the approach taken towards the Winter Smoking Room (described above), in that all issues must be considered at all times and never considered in isolation.[9]

Conservation planning

Cardiff Castle is now embarking upon a major conservation project. It is valued at approximately £8 million and is supported by a £5.7 million Heritage Lottery Fund grant.[10] It is the biggest conservation project in Wales and acknowledged by many to be the largest art conservation project of its type in the United Kingdom.

The project is, no doubt, a challenge, but not so much because of its scale, but because of the complexity and diversity of issues it raises. If the challenge is to be met, then the project needs a very firm foundation, based on understanding the site in all respects.

In order to amplify our understanding, a programme of research and analysis was devised in accordance with the conservation and management plan. The overall extent of research and analysis was divided up into units that could be delivered by appropriate experts. The Cardiff Castle Surveyor to the Fabric is responsible for determining the work to be undertaken, the drafting of detailed briefs, and making the appointments. He is also responsible for managing the whole process to ensure that the required outcomes are delivered at the appropriate time. In order to put subsequent discussion into context, consider one of our main problems.

Most castles have thick solid walls, which were obviously good for defensive purposes, but also help to keep the damp out. The theory is that when it rains on the outside, by the time the rain has reached the inside, the rain has stopped and moisture evaporates from the wall via the external face (Figure 3).

At Cardiff Castle, the walls are much thinner (Figure 4). Consequently, when it rains, the moisture has already reached the internal face or has reached such a depth within the wall as to begin to affect the internal face and interior decoration by the time the rain has stopped. Within most internal spaces, the environment is generally dry and warm, thus resulting in a low vapour pressure; lower than the building fabric. Moisture is thus drawn in through the wall from the outside.

Moisture cannot reach the internal face of the wall due to the impervious decoration applied to it. Moisture does, however, reach the underside of this decorative layer. In many areas this will cause deterioration due to factors associated with penetrating dampness. Building fabric that is almost continuously damp will, for example, become weakened and the bond with decorative layers may fail. Wet and dry cycles will also result in movement that will cause stress within the decoration, thus leading to separation. These are just two ways in which broken and otherwise defective paint finishes can occur.

All masonry contains soluble salts, but the salt content within the masonry at Cardiff Castle is high. This may be due partly to the many interventions that have taken place over decades involving the use of cement, such as grouting into the core of the walls. These salts are mobilized by moisture movement from the external to the internal wall surfaces.

The environmental conditions at the interface between the decorative layer and the supporting wall will fluctuate. This will result in the re-crystallization of the salts, a process that can very often break down

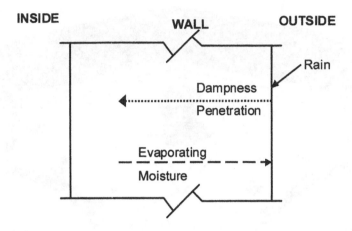

Figure 3 *Section through typical thick solid wall.*

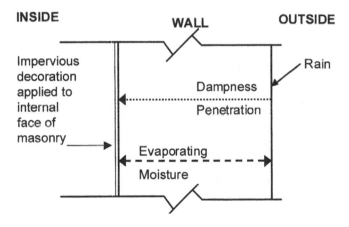

Figure 4 *Section through typical Cardiff Castle solid wall.*

the pore structure of the masonry to which the paint is bonded. The masonry erodes and the paint loses its adhesion (Figure 5). It is the failure of the paint finishes that allows moisture to evaporate and soluble salts to crystallize. This process is influenced by the environment within the building, which is affected by its current use, and also by the design, make-up, and condition of the wall and the prevailing weather.

Figure 5 *Lord Butes Study. Typical painted decoration defect: paint detaching from substrate. (Cardiff County Council)*

The research and analysis programme will study this problem with the intention of determining solutions. The programme comprised the following:

- rectified photography (used subsequently as part of the specification and for recording purposes)
- architectural paint research
- technical paint research
- environmental monitoring
- mortar and masonry
- building services engineering

The various areas of research and analysis involve different disciplines, professionals, and individuals. Regardless of how good conservation and management might be, or how efficient the project systems and procedures are, success depends upon the proficiency of the expert consultants.[11]

Appointed consultants should be appropriate to the needs of the overall research and analysis programme. This means that consultants should be happy to work in accordance with a brief and as a member of a research and analysis team.

At the outset, all research and analysis consultants must have an appreciation of the architectural paint research. This maximises the understanding of the building and the layers of decoration that have been applied.

Subsequently, some very obvious questions must be asked. What are the paint defects? What is causing them? Are the problems due to poor environmental conditions? Are problems due to moisture from the outside finding its way into the inside? These very general questions must have very detailed answers.

Proceeding with research and analysis

Each of the research and analysis consultants, together with the Surveyor to the Fabric and the Castle Keeper of Collections, has a specific role to play. In undertaking these roles, each has to liaise and sometimes work alongside one another. The roles and duties of each party, together with interactions, are described below.

Decorative surfaces consultant

1. Undertakes architectural paint research. This will assist in the understanding of the decorative schemes, the history of interventions, and the relative degrees of importance attached to each layer and component (Figure 6).
2. Takes samples of building fabric for salt analysis. These samples will be from just beneath the paint finish and from within the thickness of the wall. The latter involves working alongside the environmental monitoring consultant, who has to drill holes into the walls for the insertion of probes. The salts need to be identified for two key reasons – to determine whether the salts are hygroscopic and to establish their equilibrium relative humidity (ERH).
3. The ERH of the salt(s) present will assist in understanding the behaviour of the salts present. Table 1 gives the ERH of typical salts. Matching the salts with those listed in the table will indicate under what environmental conditions the salts will be in solution. The data from the environmental monitoring consultant will thus provide information on when the salts are in solution and when they are crystalline. The influence of these conditions is discussed under the heading of environmental monitoring consultant.

Figure 6 Small Dining Room. Architectural paint researcher (Alyson Thornton, Hirst Conservation) extracting sample for analysis. (Cardiff County Council)

Table 1 (below) Salts typically present in masonry. Illustration of equilibrium relative humidity (ERH) at given temperatures. Above these levels, salts will be in solution and thus not crystallize.

Salt	Equilibrium relative humidity (ERH) (%)						
	0°C	5°C	10°C	15°C	20°C	25°C	30°C
Calcium chloride $CaCl_2.6H_2O$	41.0	37.7	33.7	–	30.8	28.6	22.4
Magnesium chloride $MgCl_2.6H_2O$	33.7	33.6	33.5	33.3	33.1	32.8	32.4
Sodium chloride $NaCl$	75.5	75.7	75.7	75.6	75.5	75.3	75.1
Potassium chloride KCl	88.6	87.7	86.8	85.9	85.1	84.3	83.6
Calcium nitrate $Ca(NO_3)_2.4H_2O$	59.0	59.6	56.5	54.0	53.6	50.5	46.8
Magnesium nitrate $Mg(NO_3)_2.6H_2O$	60.4	58.9	57.4	55.9	54.4	52.9	51.4
Sodium nitrate $NaNO_3$	–	78.6	77.5	76.5	75.4	74.3	73.1
Potassium nitrate KNO_3	96.3	96.3	96.0	95.4	94.6	93.6	92.3
Sodium sulphate (anhydrous) Na_2SO_4	–	–	–	–	82.0	82.8	84.4
Magnesium sulphate $MgSO_4.7H_2O$	–	–	86.9	–	90.1	88.3	88.0
Sodium sulphate (hydrated) $Na_2SO_4.10H_2O$	–	–	–	95.2	93.6	91.4	87.9
Potassium sulphate K_2SO_4	98.8	98.5	98.2	97.9	97.6	97.3	97.0

Environmental monitoring consultant
1. The environmental monitoring consultant has recorded data on conditions – air temperature, relative humidity, vapour pressure, lux, and ultraviolet – within key internal spaces (e.g. rooms). Temperature and relative humidity (RH) values have been recorded at the surface and within the thickness of designated walls. External monitoring has also been undertaken, including provision of a weatherstation to record temperature, rainfall, wind speed, wind direction, and sunlight.
2. The relationship between all environments was studied, such as between a kitchen and adjacent Burges interior and between a Burges interior and the wall surface of the same room. Similarly, the relationship with the external environment and the body of the monitored walls has been established. However, such relationships can only be understood properly if the extent of dampness caused by hygroscopic salts is known (information received from the decorative surfaces consultant) and external moisture ingress is ascertained (information received from the mortar and masonry consultant).
3. Having understood the issues concerning hygroscopic salts and ERH levels, the environmental monitoring consultant determines the extent to which weather is a problem and the extent to which the use of the House is a problem. Use-related data are provided by the Castle Surveyor to the Fabric.

Mortar and masonry consultant
1. Understands the design and construction of the walls, together with the materials from which they are made and associated performance characteristics.
2. Understands the history of repair, the current state of repair, and the outstanding defects. Consultation with the decorative surfaces consultant and the environmental monitoring consultant will give an indication of where external moisture penetration occurs.

The foregoing describes, in a general sense, the tasks of the three main contributors to the research and analysis programme, together with their main interactions. At the same time, the building services engineer is getting to know the building and its problems via regular meetings with the whole team, together with inspections of the service installations.

Subject	Most suitable conditions	
	Relative humidity	Air temperature
Books and documents	Between 55% and 65% Fluctuations particularly damaging	Not higher than 15°C
Clocks and watches	Between 55% and 65% Fluctuations should be avoided	Not higher than 15°C Fluctuations should be avoided
Furniture	Between 55% and 65% with daily fluctuations less than 10%	Not higher than 15°C Avoid placing near heat sources
Mirrors	Between 55% and 65%	Not higher than 15°C
Natural history specimens	Constant 55%	Not higher than 15°C
Black and white photographs	Between 45% and 55%	Not higher than 15°C
Colour photographs	Between 30% and 45%	Constant between 8°C and 12°C (better in store between 2°C and 5°C)
Miniatures painted in water colours on ivory	Maximum 55%	Not higher than 15°C
Oil paintings in frames	Never less than 50% and not more than 65%	Not higher than 15°C
Water colour, drawings, and prints on paper	Between 55% and 65% Fluctuations should be avoided	Not higher than 15°C Fluctuations should be avoided
Parchment	Between 60% and 65% Fluctuations should be avoided	Not higher than 15°C and lower if RH control will permit
Stone (including marble and alabaster)	Between 50% and 60%	Between 5°C and 15°C
Wall paintings	Between 55% and 65%	Not higher than 15°C
Timber panelling	Between 55% and 65% with daily fluctuations less than 10%	Not higher than 15°C
Model ships	55%	Not higher than 15°C
Musical instruments	Between 50% and 60%	Not higher than 15°C

Table 2 *Ideal environmental conditions for building components and collections.*

Finding solutions to the problems

The following section summarizes what is involved in finding solutions to the problems, after all the issues have been properly considered. The principal objective is to ensure stability (i.e. keeping salts in a crystalline state). This means, as indicated earlier, keeping RH levels below the ERH levels of the salt(s) present (see Table 1). However, other building elements and furniture must be considered (Table 2).[12] It is often stated in these circumstances that a balance will be achieved. In this case, however, the balance will always be in favour of the decorative Burges interior decoration, as this has been determined to be the most important element of Cardiff Castle.

It will not always be the case that problems can be totally resolved. The mortar and masonry consultant must determine repairs and new detailing in order to minimize moisture ingress from the outside of the building. This will, in some places, be difficult to achieve.

The environmental monitoring consultant, bearing in mind all the issues raised by the decorative surfaces consultant and the mortar and masonry consultant, must determine what environmental conditions are to be achieved within an internal space in order to help bring about the optimum environmental conditions at the interface of the decoration and building fabric.

It is the building services engineer who will determine how, and to what extent, optimum environmental conditions can be achieved. Obviously, building services alone will not achieve the objectives; management and housekeeping are just as important. Here, the Castle Keeper of Collections and Curator, together with the Surveyor to the Fabric, will work with the consultants to determine appropriate solutions. These will, of course, have to consider all issues (Table 3).[13]

Appropriateness of aesthetics

The research and analysis programme will determine what conservation work will follow. In some rooms and internal spaces there may be a number of different scenarios, all of which will be technically feasible. However, as discussed previously, each of these scenarios may have a different effect on aesthetics.

Whereas dealing with technical issues can be fairly objective, aesthetics are highly subjective. Whilst the decision-making process will properly consider well-respected philosophy and good practice, the overall subject of aesthetics needs very careful consideration.

RH(%)	Typical effects
100	Saturation percentage
>96	Mould can develop on glass wool
>90	Bacteria can multiply – mould can appear on brick and painted surfaces
>85	Dampness stage – materials may become visibly damp or damp to touch Timber decay occurs
>76	Mould can develop on leather Multiplication of mites greatest above this level
>70	Viability of mould increases markedly
65	Maximum survival level for dust mites
40–50	Minimum survival level for dust mites
45	Minimal optimal comfort level for humans Electrostatic shocks more likely below this level

Table 3 *The effects that environmental conditions may have on materials and people.*

In order to assist in this, other buildings with very important Victorian interiors have been examined. The starting point has been to find out what decisions have been made and determine the basis of those decisions. Where such buildings are owned, managed, or have work that is controlled by credible and respected organizations with responsibility for the care of our heritage, this in itself gives some credibility to the approaches and treatments that have been undertaken.

The Burges interiors at Cardiff Castle can be used to illustrate many issues, and many other buildings can be used as examples to describe the possible approaches to conservation.

The Summer Smoking Room is one of the most important and the most untouched William Burges interior at Cardiff Castle. Although the research and analysis is not yet complete, conservation work will probably entail gentle surface cleaning which will make the appearance slightly brighter. The balustrading of the gallery, however, was once gilded, but is now painted. The question arises as to whether this balustrading should be re-gilded. If this were done, it would appear new in comparison with the overall appearance of the interior. Whether this causes an aesthetic imbalance must be ascertained.

Within the chamber of the House of Lords at the Palace of Westminster, it is the throne that draws the most attention. This was re-gilded about a decade ago. For those who care for the Palace, there is no question of an aesthetic imbalance here; there may very well be a symbolic reason for the re-gilding of the throne.[14] But where the balustrading of the Summer Smoking Room is concerned, there is no symbolic reason for re-gilding. These issues must be considered.

As well as aesthetic appropriateness, the setting and presentation of a room with its furniture is also a major consideration. William Burges designed many of his interiors, complete with furniture. Furnishing with original furniture is a primary objective, in line with our focus on complete interiors, as detailed in the *Cardiff Castle Conservation and Management Plan*. Thus far, this has been achieved only in certain areas (such as the Bachelor Bedroom, Summer Smoking Room, Library, Lord Bute's Bedroom, and Small Dining Room). Upon completion of conservation work to the Burges interiors, the level of furnishing will also be advanced.

Conclusion

The *Cardiff Castle Conservation and Management Plan*, by the very nature of what it is, brought together all issues concerning the site. This is not only where 'conservation (of the building) meets conservation (of the contents)', but also where economics, tourism, and other issues (including other conservation issues such as wild life and ecology) are duly considered.[15] The issues highlighted in this paper are indicative of the diversity of the professions involved.

The research and analysis programme is another area where the need to consider all issues is only too obvious. None of this happens unless appropriate management is deployed. Management, in terms of the systems and procedures that are in place and the personnel performing the management function, should ensure that everything that should happen, does happen, and at the appropriate time. This undertaking can, therefore, only be achieved successfully if the personnel concerned are well acquainted with all the key issues. Management is thus another area where all issues are considered and where 'conservation meets conservation'. It is only with appropriate management that the application of knowledge is properly applied to the overall and balanced benefit of Cardiff Castle.

Biography
John Edwards DipBldgCons(RICS), MCIOB, MRICS, IHBC
John Edwards is Cardiff Castle Surveyor to the Fabric and the client for the £8 million Cardiff Castle project, on the behalf of Cardiff Castle and Cardiff County Council. He has worked in the building industry for over 25 years, in both the public and private sectors, including building surveying, project management, and facilities management. For over half his career, John Edwards has worked with historic buildings, and has been involved with Cardiff Castle for more than ten years.

Acknowledgements
The author would like to thank the following members of the research and analysis team: Elizabeth Hirst, Karen Morrisey, Paul D'Armada, Alyson Thornton, Huw Lloyd, Dr Jagjit Singh, Professor John Ashurst, Catherine Woolfit, Graham Abrey, Matthew Williams, Martin Thomas, and Anne Philps.

Notes
1 Edwards, J., 'Conserving Cardiff Castle', *Journal of Architectural Conservation*, Vol. 8, No. 1, March 2002, p. 7.
2 Crook, J. Mordaunt, *William Burges and the High Victorian Dream*, John Murray, London (1981), pp. 263, 265, 280.
3 Edwards, *op. cit.* (2002), pp. 14–16.
4 ICOMOS, *The Venice Charter: International Charter for the Conservation and Restoration of Monuments and Sites* (1964).
5 ICOMOS, *The Burra Charter: The Australian ICOMOS Charter for the Conservation of Places of Cultural Significance* (1999).
6 Edwards, *op. cit.* (2002), p. 17.
7 Thomas, S., 'Approaches to the Treatment of Historic Painted and Decorated Interiors', *Journal of Architectural Conservation*, Vol. 3, No. 1, March 1997, pp. 19, 22–23.
8 Edwards, *op. cit.* (2002), p. 17.
9 Ibid., pp. 9–10.
10 Ibid., p. 8.
11 Ibid., pp. 20–2.
12 Sandwith, H. and Stainton, S., *The National Trust Manual of Housekeeping*, Penguin Books, London (1991).
13 Oliver, A., *Dampness in Buildings*, 2nd ed. (revised by Douglas, J. and Sterling, J.), Blackwell Science, Oxford (1997), p. 18.
14 Jardine, T., Hay, M., and Carter, S., personal communication, 2 February 2001.
15 Cardiff County Council and Ferguson Mann Architects, *Cardiff Castle Conservation and Management Plan*, unpublished report (2000).

Management of the Historic Environment – The Broad Nature of the Process

DONALD HANKEY

Abstract

The process of managing the conservation of the historic environment applies to artefacts as well as to their context and the lessons learned are common to all scales of conservation challenge. Indeed, the ethical nature of the process requires the building of consensus amongst all stakeholders involved. This can only be achieved by promoting the best scientific, social, and cultural understanding.

This paper examines how conservation can be sustainable only if all threats and weaknesses have been accounted for. While the body of scientific knowledge and technology has greatly increased, social, economic, and cultural factors remain a necessary and integral part of the equation for achieving sustainability and the support of stakeholders. The paper will try to set the scientific and technical challenges into the context of the management process.

In an age of increasing educational and professional specialization, there is a danger that policy and practice may suffer as a result of myopic perspectives and the lack of a common language to define the values and significance that must be defended. Sustainable solutions essentially involve multisectoral interests. This paper examines the common nature of the management processes that must be followed by all participants in any conservation project.

Introduction

The conference title *'Where Conservation Meets Conservation'* has neatly inferred the wide range of criteria required to achieve sustainability. It is important to recognize, however, that 'conservation science and technology' are not the only factors that underpin sustainability. I wanted, therefore, to examine the question of what constitutes sustainability, through a wider perspective. After all, if we look at only part of a problem, we inevitably get only part of the answer. The challenge is great for problems of conservation since they involve such a wide range of factors.

The concepts of 'value' and 'significance' underpin our definition of heritage through cultural, artistic, scientific, social, political, historic, and economic perceptions that may not be easily quantifiable. Yet it is these values that must be sustained in any conservation project; sustained as a measure of conservation in any necessary process of change; and sustained for the appreciation of the public as well as for the stakeholders in the conservation objectives.

In the development of a project, the public often wish to see that planning controls are meaningful; that the expenditure, perhaps from fiscal sources, is worthwhile; and that the benefits to the community are realized through enhanced education, tourism, and commerce, and through social, cultural, and political identity.

A wide range of specialist skills can be involved, and it can be a great challenge for project management to encompass the required research, planning, design, and management. Our projects depend upon skilled professionals and craftsmen with great knowledge and attention to detail. This can lead to distorted judgement on wider issues. Good teamwork to resolve all of the issues is essential to achieve the objective of sustainable conservation. This, it is suggested, requires an awareness by all involved of the opportunities, constraints, and options available, the ability to make valid choices in relation to the 'market' for future use, and the ability to realize the construction and to manage the project over time.

Management as a necessary component of sustainability

It is clear that architectural and object conservation are very similar in their use of conservation science and technology, and in the setting of values and significance to be conserved for the future. Each conservation objective will certainly require its own particular emphasis for its management and realization. But achieving sustainability of sites, buildings, their contents and of movable objects, and including consideration of socio-cultural factors, depends upon completing a range of generic management actions that relate to:

- surveying, recording, researching, understanding, and communicating (about the asset, its value and significance, and the risks and benefits), to achieve a conservation plan that should be brought forward with the project;

- realizing opportunities (associated with conservation and use, and in response to present and predictable demand);
- defining the affordable action (re-use, operation and maintenance, research, interpretation, presentation, and so on);
- assessing the costs and benefits (risks and change, revenues and funding, direct and indirect benefits, affordability);
- defining and planning the management actions (required to reach the desired objectives);
- setting out the business planning objectives (to include capital costs, operational and management budgets, funding and financing, and limiting parameters for use and conservation); and
- finally, confirming feasibility before preparing production information for any action.

We have heard from many of the contributors about the integrated nature of the conservation and management action that is required. I wanted to look, in outline at least, at the process of managing and realizing a conservation project, whether it is for an object, a building, or an area. Indeed, I am tempted to suggest that the generic actions, noted above, underlie all management in principle, not only to achieve conservation objectives, but also to resolve political and social questions, and for the generation of policy by government.

The context and need for stakeholder participation and consensus

In some of our commissions we are members of a team, and may not be in charge of all aspects of the management of the project. But the experience of Gilmore Hankey Kirke Limited (of the GHK Group of Companies) on a wide range of projects suggests that it is essential for all stakeholders to understand the enabling and team role that their particular position must play. The range of stakeholders in a project can be very wide, and their perspectives and agendas must all be taken into account if true consensus on the plan for the future is to be maximized.

Publicity about the value, significance, opportunities, and benefits associated with the social culture and with the physical heritage is essential for support to be widely based. Socio-cultural perceptions are at the heart of defining value and significance, and are often ignored by the specialist dealing with the physical environment. Our challenge today is to better understand the human need that our work on the physical environment is intended to satisfy. But social factors are often undefined or ignored, and the human expected to adapt. Post-war housing showed how a lack of such considerations can lead to failure of the social structure. In the same way, conservation that fails to account adequately for the requirements of the user will fail in its management and maintenance.

The physical heritage is more often than not protected by law, and is a public as well as a private good. This requires us to be sensitive to the community's appreciation of the project objectives, since public participation or perception can be very important factors in achieving sustainable results and good governance. Not only public policy, but in many cases public money, underwrites implementation, and is intended to achieve sustainability. But how is a conservation project to be sustainable if its value, risks, and benefits are not clearly expressed and understood? Neither the general public nor the politicians, including their enabling financial departments, are likely to give support unless the risks and benefits can be appreciated.

But it would be wrong to believe that conservation is a unique process. It is only one aspect of a range of considerations in the existing social, political, economic, and physical environment. Conservation of the existing environment lies at one extreme of all of these considerations. Furthermore, I believe that due consideration of the contribution that conservation can make is essential for the equitable management of the process of any change.

The building of consensus and participation, community involvement and ownership, social equity and stability, especially for the larger conservation projects, lies at the heart of good management and sustainability. These objectives lie also at the heart of policy for environmental upgrading by the world's leading financial and development institutions.

The influence of the sponsor and institutional structures

Success often depends upon the project sponsors and upon their understanding of the management and design processes required for achieving the objectives. But social or institutional failings can undercut delivery strategies, and appropriate and sustainable solutions may be influenced by the often limited freedom of action available to the sponsor; upon the nature of the local institutional and political systems; upon vested interests in the *status quo* and the hidden economy; upon the availability of scientific and technical skills and education required to achieve acceptable methods of handling the existing environment, or upon the availability of contracting, design, and management skills; and often upon the awareness of the benefits and risks in taking appropriate action.

There are many foreign and United Kingdom cultural assets where the conservation technology required to conserve, upgrade, and develop appropriate use is relatively simple, but where the sustainability challenge lies in achieving adequately supportive institutional and management frameworks.

Take the example of Stonehenge (Figure 1). It seems to me, and this is corroborated by the local district councils and by the Ministry of Defence, that the problem in the early years lay with the sponsorship of the design competitions of the development of Stonehenge and management of the project. Evaluation of the available options was never undertaken and consensus agreement not reached before the solution of design competition winners was thrust upon the many stakeholders. The project was thus delayed by many years. The site presents significant challenges for the restructuring of the tourism market in the region. Little assessment of the impact of tourism or potential for regional benefit has been carried out and there has been inadequate consensus reached within the local district councils on the appropriate restructuring that would be required. Yet it may be essential to assess the potential market and its impacts now in order to structure the content of the project and its benefits correctly, in relation to its context and the range of stakeholders involved.

Figure 1 *Stonehenge in its landscape setting.*

Should the Government, here or in many other projects, continue to pick up the cost of heritage where the potential direct or indirect benefits have not been adequately investigated? Without market assessment, can we be responsible for sustainability? And if we cannot prove sustainability, should our feasibility assessment stand? The Heritage Lottery Fund (HLF) would perhaps say no!

Problems are not confined to the United Kingdom. They abound where there is a lack of professional skills, institutional management and the means of building consensus, which is so often apparent in the developing world.

In 1998, the Leshan Buddha (a World Heritage Site in Sichuan Province of China, dating from the eighth century AD) was desperately short of design, management, and technical skills (Figures 2a and 2b). There was no project architect, and middle management was unable to manage without provincial and central government support. The local consultant engineers did not understand the value and significance of the site and the need to conserve the natural setting.

The project involves work to improve access, upgrade tourism facilities, and develop the presentation of the 12 sq km site. Only since UNESCO (United Nations Educational, Scientific and Cultural Organisation), the Central Chinese Government, and a leading academic from Qinghua University in Beijing have become involved has the design for new construction begun to incorporate the principles of reversibility to preserve the natural setting, and with minimum impact, to respect the conservation environment.

Figure 2a *The Leshan Buddha site seen from the Ming River.*

Figure 2b *The Leshan Buddha, at 71 m high, the largest in the world.*

The remarkable Huguang Huiguan complex of guildhalls in Chongqing China, dating from the eighteenth century, at first lacked political organization and support from the local government (Figure 3). The local university was the only organization that could attempt the planning of the restoration, but it lacked the technical, scientific, and management experience to undertake the work. Two years were lost in achieving institutional and political consensus, appropriate relationships between the stakeholders, and in developing the necessary skills, and achieving support for the benefits of the project. The project required the local district government to develop a new vision for local urban planning, adjacent land use and urban infrastructure in this central part of the city. All of the lessons for understanding the market, defining appropriate use, and for business and management planning still have to be learned. These concepts are hard enough to achieve in our democracy, let alone in the more disenfranchised environments of centralized government.

The walled city of Lahore in India was on the point of irreversible decay in 1987 (Figure 4). As we in GHK found in Calcutta in 1996, the city needed to carry out the basic management processes: record its heritage, evaluate its existing historic environments, and develop strategic and

Figure 3 *Part of the roofscape of Huguang Huiguan in its present urban context.*

Figure 4 *Lahore, street market encroachments showing a lack of city planning and development control.*

structural planning policies and catalytic projects, so as to encourage investment in conservation and re-use. The skills of the local consultants needed strengthening, and the administrative and management methods of the local city offices needed the benefit of sound experience, integrity, and good practice. The City required the rationalization of appropriate construction and new technology for re-use and upgrading, and improved institutional and management structures to achieve sustainable solutions.

The scientific and technological requirements for environmental upgrading and re-use were important challenges to the city administration. The project sponsorship and management, planning and architectural direction needed strengthening. Business planning, in relation to the market opportunities, and institutional support and vision had been lacking. It was such factors in these large projects that were the most significant in threatening sustainability.

The universal management process

ICOMOS-UK is one of the present 110 member countries of the International Council on Monuments and Sites (ICOMOS). It is a membership organization responsible for the management plans of world heritage sites and is supported in part by the Department of Culture, Media and Sport (DCMS). It has specialist sub-committees for conservation research and technology, tourism, architectural and planning subjects. It has a voluntary membership of skilled professionals and concerned individuals. One of its sub-committees has investigated and published, in draft, a document entitled 'The Management of the Historic Environment'[1] When amended with the benefit of the comments from member countries and the many United Kingdom stakeholders, the draft will be presented as an ICOMOS international guideline document. It sets out to cover for all cultures the most generic management principles for the historic environment.

The general comments given above form some of the context within which management must achieve sustainability and a measure of good practice. In the process of good management there is a place for all of the subject matter of this conference. In presenting this conference paper, I wish to emphasize and, at times, illustrate some of the important points that the ICOMOS draft document contains:

• The guidelines apply to 'all those involved in decision making for the historic environment', including the stakeholders (such as politicians, administrators, owners and site managers, the local community, and other interest groups) and professional advisers (such as historians and archaeologists, land-use planners, infrastructure engineers, sociologists and architects, craftsmen, economists and financial analysts). A simple flow chart of the main steps is given in Figure 5.

- Understanding the values and significance of the historic environment is essential if we are to conserve that significance, and if we are also to manage the processes of change in ways that maximize benefit and achieve sustainability. Interpretation, presentation, and education promote such understanding.

- All management strategies and plans should be economically viable, widely accepted, and achievable within specified time scales in the context of the available technical and financial resources.

- Sustainability of the historic environment will be improved if all the stakeholders have the opportunity to learn of its value and significance; if they can understand the physical, social, financial, and economic benefits and constraints; and if they are able to give active support towards funding the necessary operation and maintenance.

- Whoever leads should have an understanding of the range of skills required for the management, planning, design, construction, operation, maintenance, financing, and funding of the historic environment.

- The development of a strategy is essential for successful management of the historic environment. It ensures that conservation and economic issues are considered, that all stakeholders can be involved in the determination of its future, and that options for the future can be properly considered.

The following logical process serves as a guide. The first six steps – (a) to (f) – could form a separate conservation plan or they could be the first part of a comprehensive management strategy. Ideally, these steps should be completed before any detailed consideration of the preferred development option for a site or object.

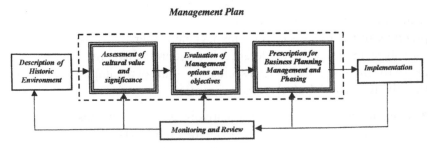

Figure 5 A strategy for the management of the historic environment.

(a) Value and significance

Description of the site, leading to the identification and assessment of its cultural significance, and of the values of the different interest groups upon which the significance is based.

The listing designation of buildings and conservation areas has encouraged the recognition of urban grain, and has encouraged determination of the values and significance of the historic environment. English Heritage's listing, as for most countries, has so far shied away from assigning values and significance. Such matters remain to be researched by the building owner's team. The research, recording, and assessment of value and significance are fundamental to the start of any conservation process, and underwrite all subsequent consideration for project management.

(b) Condition of the asset

An assessment has to be made of the condition of the asset and its ability to adapt without undue loss of value and significance.

In looking at the technical concerns for conservation, and because of today's requirement for improved servicing and environmental standards, we are working to stricter scientific margins for conservation and the impacts of future use in upgrading historic environments; for the use of green materials; for energy saving; for functional, economic, and physical sustainability; and for returns on investment. We seek tangible benefits to justify the intended capital works, operation, and maintenance of the project. These stricter requirements apply to all adaptation, as well as to all new building. They apply to all historic environments and, in similar ways and to different degrees, to the artefacts that they contain.

(c) Pressures affecting the site or object

An assessment has to be made of the existing pressures affecting the site or object, its management context, user groups, stakeholder interests, legal controls that affect its significance, and other forces for change.

External economic and physical pressures can impact heavily on the historic environment and on their contents, functions, and historic design. These pressures form the context within which sustainability must be achieved. For instance, the historic city of Li-Jiang in Yunnan Province has, since 1999, been a Chinese World Heritage Site. It was seriously damaged in an earthquake in February 1996, causing 400 deaths in the region and 4,000 homeless (Figure 6). The traditional wooden buildings around domestic-scale courtyards, and the new concrete structures and

Figure 6 *Li-Jiang after the earthquake.*

infrastructure services, required extensive conservation and recon-
struction. The city took the opportunity given by the earthquake to
remove incongruous structures.

There is still, however, a great need to validate the conservation and
repair through ensuring the social, economic, and functional sustainability of the traditional urban environment. In the short term, the attrac-
tion of the historic environment has brought many visitors and traders.
But in the long term, it will be essential that the economic category of
resident is able to maintain and upgrade the quality of the traditional
environment, retaining the architectural and historic detail, and its values
and significance in the life of the local population. Upgrading the historic
structures and environments is essential to retain the status of the old
city in the market place. This demands technical and scientific solutions,
the creation of a design and planning grammar, and construction guidelines backed by development control and urban conservation, within
which appropriate change can occur.

The conservation danger here is that the richer resident moves out to
the adjacent new city with modern services, and the poor move in leaving
a decaying and sub-standard historic, but overcrowded, environment.
In historic sites and cities in the United Kingdom and all over the world,

the potential for conservation is closely linked to external demographic, economic, and land use pressures.

(d) A statement of the conservation policy

The governing principle of the conservation policy is the need to protect and enhance the significance of the site. Limits of potential change should be identified that will cause the least loss of value and significance, while also ensuring sustainability. The policy should be comprehensive and cover all significant aspects of the site. It should be understood by all the stakeholders and should be accepted and adopted by the relevant planning authorities, where applicable.

(e) Evaluation of the options

An outline evaluation of the options for managing the pressures affecting the significance of the site involves the evaluation of possible adaptation for future use, and the identification of the preferred option for each policy area.

Take, for example, the city of Shibam in Yemen, a World Heritage Site of mud buildings constructed some eight storeys in height (Figure 7). The participation of the people in the conservation process is absolutely essential if the buildings are to continue in use. As mud disintegrates when

Figure 7 *Shibam from Wadi Hadramut showing buildings of eight storeys in height using pure mud and straw traditional construction, showing on the left the results of failure to install water and drainage services correctly.*

wet, the introduction of upgraded and serviced environments in a manner that conserves the value and significance of the historic and cultural environment is a considerable challenge. It involves the co-operation of all levels of the administration and the community. It requires setting standards and limits for construction and maintenance methodologies and communicating them to a largely illiterate population. In determining the options for future use, the conservation process requires the understanding and co-operation of all stakeholders, and each household must follow traditional and often innovative construction methods for installing services that will not damage the structures, carrying out repairs, and renewing finishes on roofs and walls especially prior to the seasonal rains. There is an intimate relationship between the people, the architecture, the contents, the economic and financial nature of the functions, and the required management and maintenance disciplines required to achieve a sustainable historic environment.

In the same way, historic buildings in the United Kingdom, made of uninsulated materials and sometimes with fine internal finishes, are unable to accommodate the adaptation required to upgrade their environments to the demands of the twenty-first century. In order not to damage the historic fabric, limits must be imposed on the upgrading of heating systems, the addition of insulation, the installation of electrical services, and on the location of any change that may have potentially damaging (such as damp) impact.

In the case of quite simple structures, there is a need for great expert understanding of the liabilities and conservation requirements of the existing fabric, the options available for the upgrading and installation of services, and of the options available for future use, without endangering the historic fabric. The role of the conservation consultant may need to be quite prescriptive. There is certainly a need to understand values and significance to an extent that we can make rational decisions concerning the necessary compromise and adaptation required to achieve economic, financial, and physical sustainability.

(f) Market assessment

A critical step in assessing the extent and appropriateness of potential adaptation and use is the evaluation of the demand for future use; the financial, cultural, social, and environmental character of the market; and of the potential forces for change. Such pressures have to be balanced with the potential of the fabric to be conserved and to adapt.

(g) Planning for overall management, business plans, operation and maintenance

Policies for the conservation and enhancement of the existing environment bring manifold economic, financial, social, and other benefits. Through recent experience both in the UK and abroad, we are beginning to understand the tangible and intangible benefits that accrue from different types of project. But there is still much to understand in this field, and it is vital – for political, financial, and economic support – that the benefits of the conservation-led schemes are predictable. The rules of all good business apply equally to the heritage, and we have to ensure that the cultural property is not at risk and unsustainable from the wrong decisions being taken. The HLF, in using public funds, has had a beneficial influence in demanding the evaluation of business plans, of requiring the likely benefit and sustainability to be stated, and in demanding good management plans.

Making early sense of the financial frameworks and limits for change, and testing the potential direct and indirect benefits of the favoured options for change, is fundamental to selecting the sustainable solution. This requires the adequate evaluation of budgets and costs for capital works, operation and maintenance, and defining the sources of revenue. It requires the drawing up of management and business plans; the balancing of possible expenditure with revenues; the assessment of likely benefits, risks, and constraints; and the assessment of the need for any permanent outside funding, perhaps through the fiscal system, on the basis of the beneficial indirect impacts of the heritage on the local economy. Only then can we select and confirm the feasibility of the preferred strategy and option for change.

In effect, the above steps – (a) to (g) – should be achieved in outline for the feasibility study and any detail left for the pre-contract stage.

(h) Implementation

Identification of the ways in which the plan will be implemented and by whom, and the appointment of staff with appropriate experience and training, needs to have been considered in the management and business plans, but worked up in detail as contracting stages approach.

(i) Monitoring and review

Identification of measures for monitoring and review of the plan are required. This should be carried out as a permanent component of the management process in the years to come.

(j) Research and records

Definition of the required research and records is important for the heritage, and disciplines for the registration and archiving of all data that can assist future management and research into the cultural resource must be considered. Much money is wasted by building owners and project sponsors because work is repeated through the lack of disciplined record keeping, and further intelligent interpretation is also threatened. These actions need to be built into the budget, and into the professional and management disciplines of any project preparation.

(k) Final project

The final project requires the development of details for the works; the market and economic context; management structures; budgets, costs, and business plans; balancing of expenditure with revenues; and programmes for future action.

(l) Final statement of the vision

This requires a detailed project plan and a detailed prescription for the management, operation, and phasing of change, with the identification of a long-term vision, medium-term targets, and annual work programmes. Programmes will need to be prioritized in accordance with available resources.

Conclusion

I conclude with the thought that sustainability of the conservation process can only be achieved if the content and context are assessed in their totality. I maintain that the management process of both artefacts and sites follows a remarkably similar pattern. I also maintain that there is value in adopting a common discipline as set out in the ICOMOS draft guidelines. Furthermore, in all of the above, the United Kingdom has great talent and experience: in good management; science and technology; social and political equity; and efficient, transparent and accountable institutional, legal, and administrative systems. I would encourage you to see that these talents are exportable, especially in relation to the conservation and protection of the existing environment, and can provide a great resource for participation and development of today's world, and in a manner that promotes respect for our many cultural differences.

Biography

Donald Hankey RIBA

Lord Hankey is founder of Gilmore Hankey Kirke Limited, architects, now part of the GHK Group of Companies, which take account also of planning, engineering, economics, management and sociology. He is architect to the Dioceses of Southwark, Oxfordshire, and Bath and Wells, and to English Heritage. Lord Hankey is also consultant for international agencies for conservation architecture and conservation planning, urban upgrading, and institutional and administrative strengthening in Pakistan, Barbados, Yemen, Tunisia, India, Russia, Nepal. and China. He is Vice-President of ICOMOS-UK and Chairman of the Research and Recording Committee; Vice-Chairman of the Historic Buildings Advisory Group for the Ministry of Defence; and founder of the All Party Group on Architecture and Planning in Parliament.

Notes

1 ICOMOS, *The Management of the Historic Environment*, draft consultation document, printed in 2002 by ICOMOS-UK, (10 Barley Mow Passage, London, W4 4PH). This is now out for consultation with the 110 country members of ICOMOS. Copies of the draft may be obtained from ICOMOS-UK and comments would be much appreciated.

Discussion session

CHAIRED BY DONALD HANKEY AND
NICHOLAS STANLEY-PRICE

Delegate *(not identified)*
Is there, in the experience of the panel, a feeling of escalation in the deterioration of cultural property? During a recent visit to Malta, and in relation to limestone conservation, there was a general consensus that deterioration had escalated immeasurably over the past 50 years. In thinking about the objects that I deal with, it is felt that there is an escalation in deterioration.

May Cassar *(University College London)*
In relation to Malta, there is increasing awareness that accounts for what probably is a perceived escalation in deterioration. There are also environmental problems on what is a relatively small island, including issues relating to environmental impacts, such as industry and pollution, which are having an increasing effect. Climate change may also be having an effect, with a perception of less, but more intense, rainfall.

In general terms, I have a sense from looking at environmental data and talking with people that there is a feeling of a reduction in resources to keep up basic maintenance, causing concern for how we are going to cope in the future. In talking about climate change, the general response is that it is another factor that people do not wish to hear about.

Nicholas Stanley-Price *(ICCROM)*
I agree entirely with the two factors highlighted by May [Cassar], probably placing more emphasis on the fact that we are much more aware of the cultural heritage, with more people looking at it and evaluating it, and with greater awareness of the condition of things. Taking up a point made by Tobit [Curteis] in his paper, impressions can be very misleading and unless we can document changing conditions through the use of historic documentation (which remains as important as ever) or through systematic analyses, we cannot justify our observations on the rate of deterioration. By way of example, I am aware that one can be very impressed when visiting a site for the first time. On the second visit it can appear to have deteriorated badly since first seen. The answer is

that you did not notice the deterioration the first time, as you were so impressed with the site.

Tobit Curteis (Tobit Curteis Associates)
Rather an unpopular comment, but we should not forget that 85% of wall paintings that we treat have been treated by conservators or restorers in the past 50 years. We must remember that what we do has a tremendous impact and the growth of conservation has good and bad effects.

Helen Hughes (English Heritage)
I agree with Tobit Curteis's last point. It is important to educate the people with responsibility for the day-to-day cleaning and running of buildings. Such people can alert us to problems of deterioration. Also, all information provided by conservators should be passed on to those use and live in the buildings. We are terrible at communicating down to the grass-roots level and these people can alert us if something is going wrong. If we take the trouble to explain why a particular surface is important, then they would also know why it is important and so care for it, rather than being imposed on by people like us. There is a need for very basic information and to share knowledge, as many points (such as about relative humidity) can be communicated in simple terms. It is not rocket science. We must learn to communicate better.

Mu-Shan Chen (Yun-Lin University, Taiwan)
I have established a graduate school in conservation science at Taiwan University, but am concerned about training conservators in conservation management. I have come to the United Kingdom to learn how to promote conservation science education. Can the panel explain how to promote conservation science and how to balance this with good management?

Nicholas Stanley-Price
First, in Taiwan, you have the opportunity to set up a new programme and learn from the experience of the different programmes running in this country and elsewhere. You have the chance to benefit from interdisciplinary co-operation in designing a new programme and the opportunity to cross interdisciplinary boundaries. Secondly, and on a different point, we should look at how conservation has changed. There is a much greater awareness of the relative value of different approaches to conservation, depending on people's values. The idea of universal standard setting, seen in conservation at an international level during

the 1950s and 1960s, is now looked at in a much more relative fashion, especially with regard to the differences in philosophy between east and west. These are areas in which we have learned most and which there is much more flexibility. There is an opportunity to build these into a new programme.

Helen Hughes

You are obviously involved in commissioning different experts in projects. Are there mechanisms for allowing these experts to talk to one another? In conservation, there is often a problem where the project leader draws all the information together and then holds on to it. It is important for experts to get to talk to one another. How can we ensure that this happens?

Sarah Staniforth (The National Trust)

Absolutely. We have noticed a type of divide and rule, when the managers of projects would consult different experts, and indeed sometimes consult more than one expert from the same discipline, until they got the answer that they wanted and looked to cost the least. That process was extremely unsatisfactory for the professionals involved. The system of project management that The National Trust now uses is to have a project manager in place (for the larger projects), who will run a team of different experts according to the project. These experts will meet on a regular basis and then bounce ideas off each other.

Katy Lithgow (The National Trust)

There are often changing standards in technology, which can pose a problem. A project management system has to identify the things that it can identify, which then gives capacity to deal with the things that it cannot.

Donald Hankey (Gilmore Hankey Kirke Limited)

A project manager should take it upon himself to make sure that there is consultation between all the members of the team, who are all stakeholders in the project in their own way. It does not matter if such a person is a planner of a city, an architect, a designer for an exhibition, or an individual working on a small problem; the principle, I maintain, must be for all parties to be brought together and fused into a team. It is absolutely critical that everyone is brought in early to avoid the problem that Helen Hughes raised.

Helen Hughes

In commissioning new research, and discovering new information, this has to be fed into the existing team and not held on to until it is published. Everyone must be very humble and share information.

Michael Morrison (Purcell Miller Tritton Architects)

I get the feeling that you are working in a very different world from myself, where it is common to be screwed down to a fixed fee. On a modest project, I can honestly say that we have spent the entire allocated fixed fee on meetings, consulting all the various parties, before single thing has happened on site. It is a very expensive and time-consuming process if you are going to consult everyone. I am not saying that it should not be done, but there are practical reasons why it does not happen.

Donald Hankey

There are key ways of making sure that everybody is participating and talking to each other without doing all the work yourself.

Helen Hughes

An e-mail chart, which is constantly updated and sent to all parties, is an economical way of doing that.

John Edwards (Cardiff County Council)

The way that research and analysis have been organized at Cardiff Castle over a 12-month period has been for all consultants to meet on specific days or sets of days, over lunch, to allow the transfer of information without using up valuable time. Also, the job of project manager is to state the obvious, as it is not always obvious to everyone. Stating the obvious is the right thing to do. Producing reports just to transfer information is not worthwhile, so e-mails sent to all parties is a better way to share information. The key is to choose the right people, who want to work together and share knowledge, and get enthused about the whole project and not just their own little bit of it. It is all about selecting the right team and getting simple and straightforward processes in place. Nothing too complicated; if too complicated, it will just fall down.

Tobit Curteis

I am rather more on Michael Morrison's side on this one. I think the trouble with large and complex projects is the amount of information that is produced. To read and understand fully the number of written reports, and make sense of them, takes a great deal of time. On large projects, there are large reports coming through all the time. The key to success is if the project manager is in a position to send the relevant information to the relevant person. This does not mean that people should be out of the loop, but if all reports are sent to all persons, they are just not read. It is down to the project manager to know exactly who should have which bit of information.

Delegate (not identified)
It is possibly a good idea to produce an index and circulate this on a regular basis and so give people the option on whether to receive the reports.

David Watt (De Montfort University)
Given the broad and increasingly complex range of skills required in practical conservation, are we looking at the need for a clearer inter-disciplinary training of professionals? Are we actually looking for a new breed of conservation professional?

May Cassar
We are about to launch a new interdisciplinary master's programme next year, and what we aim to do is attract architects, engineers, conservators, and conservation scientists to the course. What we are not intending to do is produce a new breed of professional. It is an enhancement of existing skills of all the professionals that want to take the course a better understanding of how they might integrate with other professionals than to try to produce somebody new. The intention is to actually enhance skills and develop a sensitivity to the contributions that other professionals of the team actually make to any complex project.

Nicholas Stanley-Price
I agree. I think the emphasis should be on communication amongst specialists. The question of how to work with other professionals has to be directed at people who already have some professional expertise in some field. All the speakers and members of the audience have professional expertise in some discipline. When faced with the realities of the workplace, they realise that being able to communicate outside their own discipline is necessary. That should come at master's level or through continuing education.

John Edwards

I am not sure that I agree, because, as far as I am concerned, everybody should have some understanding of historic buildings at undergraduate level because so many of these people, like us, get involved with historic buildings and have the potential to do so much damage. Some understanding at undergraduate level is essential. It is, after all, and in many ways, purely maintenance.

Robert Wilmot (Historic Scotland)

Michael [Morrison], I am rather intrigued to know whether there are conservation plans at Knowle [National Trust property in Kent] and Shackleton's hut [Sir Ernest Shackleton, 1874–1922, British Antarctic explorer], as many of the things that you discussed would be prime fodder for such studies?

Michael Morrison

The conservation plan for Shackleton's hut is being written at the moment, so things are going backwards and forwards via video conferencing and e-mails. This is exactly the issue that is taking place at this intriguing place, although it is back to front as there has been quite a considerable amount of conservation, at the level of conserving little objects, without this overarching view of things, which I am now hoping to replace.

At Knowle, I am rather shamefaced to admit that there is no conservation plan, and I think we all feel that there should be. It is partly because it is an extremely difficult property, with partial occupation by the family. This means, with elderly members of the family in residence, that access to the rest of the building is not always easy to obtain.

Sarah Staniforth

Can I just say, on behalf of The National Trust, that we realise that we are deficient in not having conservation plans for all of our properties, and part of our national strategic plan (which is from 2001 to 2004) is that by February 2004, when that plan expires, every property will have a conservation plan. I do, however, have to point out that we have thousands under thousands of properties – be they humble vernacular buildings to something as complex as Knowle. We have to have conservation plans for complex buildings like Hardwick Hall [National Trust property in Derbyshire], but conservation planning is a relatively new exercise and, just like everything else we do, we learn as we go along. I also think that Knowle may actually benefit positively from being of the later, very complex, properties to have a conservation plan.

Michael Morrison
There are an awfully large number of bad conservation plans out there, and I wonder why people admire them.

Donald Hankey
May I add that there was a vogue with everybody being required to have conservation plans when, in my humble opinion, I think that everybody requires management plans, which are built up on conservation plans and are one of the elements you are going to defend in developing the future of your asset. An asset or a liability – it still requires a management plan, and perhaps a business plan, to get the sustainability.

Mu-Shan Chen
In Taiwan, most projects are not successful because the co-ordination of the team is difficult. This is the challenge of cultural property. Education should be from primary school to promote conservation science. I would be interested to know how to pass the experience of the United Kingdom to Taiwan and Mainland China. In Japan, conservation science teamwork is very successful.

Donald Hankey
In China, where I have worked extensively over the past ten years, I believe that you have to start at the top, and the Ministry of Construction and the state administration for cultural heritage are probably the agencies that have to be convinced of a policy direction. Then, I think, the provinces and municipalities will follow. It is difficult for the provinces to take the initiative, but where they have done so, they have produced very good results. They have not, however, done so with the initial promotion of the Chinese government, but as a result of the extraordinary qualities of the places in which they are found.

But, really, you have a quarter of the world's population with inadequate supporting technical and management facilities. I have the task at the moment of trying to write an urban cultural heritage strategy review for China and skills training is one of the important factors that must seep down into every province and community, so that people have the right awareness, attitude, technology, and skill for carrying out the conservation process.

Conservation is only one element in the whole consideration of the environment. Fifty per cent, somebody said, of work in this country is adapting the existing environment. Whether it is Knowle or Cardiff Castle – great places that they are – they are easy by comparison with that ambivalent half-light world of somewhere that has a social and historic context (like a historic Chinese city) where the administration does not have technology and does not understand the method or management or value or significance. The answer is that they bulldoze it! And China is fast becoming a country of a thousand similar cities for that very reason. China, itself, therefore needs to learn how to go about these issues. As she is going to be the host for the ICOMOS general assembly in 2005 and the world heritage convention is going to sit there in 2003, there is considerable pressing for China to get its act together.

Sarah Norcross-Robinson (Norfolk County Council Museums Service)

More of a comment than a question. I will be going back to my County Council feeling that you have been preaching to the converted and that many of many colleagues whom I work with on these types of projects are not here despite the best efforts of the conference organizer. The buildings in our care are all historic, of one type or another, and looked after by a local authority. The main problem is that this is the first thing that is cut, not being a statutory requirement (such as education). Maintenance is the first thing to go and so we are now looking at seriously run-down buildings that have not been maintained for many years. Money has only been spent on incidents, such as a ceiling falling down, and there we are running round putting in applications for major funding for large conservation projects. Everything seems to be the wrong way around. We are not spending money on preventive works, but instead getting funds to put things right after all the damage has been done.

May Cassar

I do not know if this is of any help, but I was involved with a project about three or four years ago for Worcester City Council (I do not think they would mind me mentioning them by name as it was a useful and very positive exercise that they did). It was to do with re-housing, from poor storage, the museum collections for the whole city. The museum service was being offered by the Council a series of five poor-quality buildings to re-house the collections. The collections were scattered around 13 sites and they really wanted to bring them together on one site and to keep them in secure and stable storage. But they were up against it as the offer of the buildings was not what they wanted; they felt that they were having to accept poor-quality buildings.

They did a cost-benefit analysis on all five buildings, including an analysis on a new-build option. The new-build was going to be the third most expensive. The most expensive was the conversion of a former hospital, which would have allowed them to do everything that they wanted with the collection including educational facilities, it was too big for them and cost a significant amount of money (approximately £5m). The second option, in terms of benefits, was a new-build costing £700,000. They would never have got a new building if they had not gone through the process of actually looking at the five buildings plus the possibility of a new-build, and actually working with the Council and the finance department in assessing the costs and benefits. What came to the top of the pile (cost envelope) was the hospital (which everyone agreed was too expensive) and then the second option (which everyone thought was the best option) was in fact a new-build. Had they gone straight to the Council with the idea of a new-build, they would have been told to accept the water-pumping station whose basement flooded every other day because that was on the cards and quite likely the one they were going to be forced to take as nobody else wanted it. So it might be worth really looking at a cost-benefit analysis and seeing whether the costs and benefits of maintenance are actually better than doing major conservation projects every five or ten years.

David Watt

In response to Sarah [Norcross-Robinson], having worked as a conservation officer for the same local authority for a number of years, I can attest to the amount of fragmentation within an authority of that size. Where you have a direct service organization undertaking repair and maintenance of historic buildings, and there is a conservation team within the same department and the two do not speak one with the other, you can begin to see the difficulties. I do think that, within such organizations, the skills are there and one needs to build bridges.

Delegate List

Irene Ampatzioglou	De Montfort University
Miranda Atkins	RICS Foundation
Helen Axworthy	R.H. Partnership Architects Ltd
Jenny Band	Hampton Court Palace
Jan Birksted	De Montfort University
Graham Black	R.H. Partnership Architects Ltd
Louise Bradshaw	Unknown
Janet Brookes	The National Trust
Carol Brown	Historic Scotland
David Brown	De Montfort University
Catharine Bull	Purcell Miller Tritton Architects
Anna Bülow	De Montfort University
May Cassar	University College London
John Castleman	Norman & Underwood Ltd
Mu-Shan Chen	Yun-Lin University, Taiwan
Paul Coleman	The National Trust
Belinda Colston	De Montfort University
Jim Crane	C.R. Crane & Son Ltd
Tobit Curteis	Tobit Curteis Associates
Audrey Dakin	Historic Scotland
Paul D'Armada	Hirst Conservation
John Edwards	Cardiff County Council
Reg Ellis	Reg Ellis & Associates
David Farrington	Derbyshire County Council
Libby Finney	Doncaster Museum & Art Gallery
Susan Friend	Dean & Chapter of Lincoln
Christa Gerdwilker	Historic Scotland
David Gibson	David Gibson Architects
Robert Gowing	English Heritage
Mary Greenacre	The National Trust
Donald Hankey	Gilmore Hankey Kirke Ltd
Mary Hardy	Getty Conservation Institute
David Hargreaves	H-H Heritage
Richard Harris	Weald & Downland Open Air Museum

Elizabeth Hirst	Hirst Conservation
Helen Hughes	English Heritage
Pat Jackson	West Dean College
Gareth James	National Museums & Galleries of Wales
Barry Knight	English Heritage
Eloy Koldeweij	Netherlands Department for Conservation
Karen Lanes	Hirst Conservation
Richard Lithgow	The Perry Lithgow Partnership
Katy Lithgow	The National Trust
Gerard Lynch	De Montfort University
Peter Martindale	Unknown
Simon Matty	Resource: The Council for Museums, Archives and Libraries
Morag Mccormick	Oxford Brookes University
Allyson McDermott	Allyson McDermott Conservation Consultants
Michael Morrison	Purcell Miller Tritton Architects
Deidre Mulley	Independent
Gill Nason	The National Trust
Sarah Norcross-Robinson	Norfolk County Council
Anthony Parker	The British Library
Mark Perry	The Perry Lithgow Partnership
Janet Rees	National Museums & Galleries of Wales
Jane Rowlands	Glasgow City Council
James Simpson	Simpson and Brown Architects
Lynsay Shephard	Hampton Court Palace
Nigel Slater	The National Trust
Claire Smith	English Heritage
Angela Sparling	Norman & Underwood Ltd
Jane Spooner	Council for the Care of Churches
Sarah Staniforth	The National Trust
Bethan Stanley	English Heritage
Nicholas Stanley-Price	ICCROM, Rome
Alison Stooshnov	Glasgow City Council
Paul Thomas	Heritage Conservation Consultant
Alison Thornton	Hirst Conservation

Alison Walker   The British Library
David Watt   De Montfort University
Robert Wilmot   Historic Scotland

Printed and bound by CPI Group (UK) Ltd, Croydon, CR0 4YY

23/10/2024

01777695-0007